Compendium of Rice Diseases

Edited by

Robert K. Webster
Pamela S. Gunnell
University of California, Davis

APS PRESS
The American Phytopathological Society

Financial Sponsor

Busch Agricultural Resources, Inc.

Front cover photograph by Robert K. Webster
Back cover photographs by Duane S. Mikkelsen

Library of Congress Catalog Card Number: 92-71949
International Standard Book Number: 0-89054-126-4

Printed in the United States of America on acid-free paper

The American Phytopathological Society
3340 Pilot Knob Road
St. Paul, Minnesota 55121-2097, USA

Preface

This compendium was prepared for all those interested in rice and the diseases that affect it. It is hoped that the technical descriptions and illustrations will assist in identification, further study, and ultimately better management and control of the maladies that limit the availability of rice as a food staple for millions of the world's population.

Because rice is grown and studied under diverse climates and cultural conditions and in many countries, the literature is broad, diverse, and in some cases scattered and difficult to obtain and translate. We have attempted to ensure that the most recent information is included but fully realize that some may have been overlooked. Since much of the existing knowledge on rice diseases is unavailable to many rice researchers in the world, a major purpose of the compendium is to stimulate further communication among rice pathologists, which should eventually lead to an improved product.

A broad array of diseases affects all parts of the rice plant and often places major restraints on rice production. Certain diseases are limited in occurrence because of the absence of environmental conditions favoring pathogen establishment and disease development, whereas others, the viruses and mycoplasmalike organisms, are limited by the absence of the vectors. In these days of easy travel and exchange, we also hope the compendium will assist in recognition of diseases unfamiliar to some so as to enhance vigilance and efforts to prevent the introduction of problems where they don't presently occur.

Rice pathologists throughout the world owe a great debt to S. H. Ou. His book *Rice Diseases*, now in its second edition, is the most complete work available on the subject, and we hope this compendium will both complement and add to it.

Contributing authors are mentioned at the ends of sections they prepared. We thank them for their efforts and willingness to work with us on this compendium. We also thank those who allowed us to reproduce various items and those who contributed slides for use as color plates. Acknowledgment of figures and plates is given only where they were not provided by the authors of the accompanying text.

The compendium was organized and planned by R. K. Webster and M. C. Rush. K. J. Leonard served as the primary reviewer, G. E. Templeton and W. B. Bockus reviewed the entire manuscript, and M. A. Marchetti reviewed portions of the manuscript. We thank them for their insights and helpful suggestions.

Authors

Sang-Won Ahn
International Rice Research Institute
Manila, Philippines

J. Mike Bonman
International Rice Research Institute
Manila, Philippines

D. Marlin Brandon
California Rice Research Station
Biggs, California

Donald E. Groth
Rice Research Station
Crowley, Louisiana

Pamela S. Gunnell
Department of Plant Pathology
University of California
Davis, California

Hiroyuki Hibino
National Agricultural Research Center
Kannondai, Tsukuba
Japan

Clayton A. Hollier
Louisiana Cooperative Extension Service
Baton Rouge, Louisiana

Fleet N. Lee
University of Arkansas Rice Research Center
Experiment Station
Stuttgart, Arkansas

T. W. Mew
International Rice Research Institute
Manila, Philippines

Jean Claude Prot
International Rice Research Institute
Manila, Philippines

M. Charles Rush
Department of Plant Pathology
Louisiana State University
Baton Rouge, Louisiana

Raymond W. Schneider
Department of Plant Pathology
Louisiana State University
Baton Rouge, Louisiana

Robert K. Webster
Department of Plant Pathology
University of California
Davis, California

Glenn Whitney
Texas Agricultural Experiment Station
Beaumont, Texas

Contents

Introduction

In a Balinese legend, the lord Vishnu, male god of fertility and water, comes to earth to provide better food for the people who had only sugarcane juice as food. Vishnu made Mother Earth give birth to rice and then fought Indra, lord of the heavens, to force him to teach men to grow rice. Thus rice, as a source of life and wealth and as a gift from the gods, was born from a union of the divine creative forces represented in earth and water.

S. K. DeDatta

Rice as a food provides a major source of calories for a large percentage of the world's population, particularly in Asia, where more than 90% of all rice is grown and consumed by about 60% of the world's people. It is also a staple food in Latin America, parts of the Middle East, and Africa. Today, rice is produced on about 10% of all cropland (144 million hectares) in over 110 countries, ranking second only to wheat in hectares harvested. As a food crop, rice provides more calories per hectare for human consumption than wheat.

The extreme variability and adaptability of rice to different geographic locations, soil types, and environments make it, in my view, the world's most versatile crop. Although most rice is produced in tropical lowland areas, the crop flourishes in a broad range of climates from tropical to temperate, as far south as New South Wales (latitude 35° south) and as far north as Czechoslovakia (latitude 50° north), and from sea level in deltas to elevations of 3,000 m in the Himalayas (Fig. 1). Furthermore, rice is grown successfully in a broad range of water regimes (Fig. 2)—unsubmerged, or upland rice (10% of total cultivation), moderately submerged lowland rice (irrigated, 45%, or rainfed, 30%), and submerged rice, in up to 6 m of water (11%) or floating (4%).

Rice belongs to the genus *Oryza*. Authors differ in opinion as to the number of species, but at least 20 are recognized, including the cultivated rices *O. sativa* L. (Asian) and *O. glaberrima* Steudel (African). Evidence of the early evolution of cultivated rice is fragmentary and inconclusive. Some believe the Indian subcontinent to be the ancestral home of *O. sativa*. Recorded history of the cultivation of *O. sativa* goes back to 3000 B.C. in China, and there is evidence that a large portion of people in South and Southeast Asia were dependent on rice as a staple food as early as 5000 B.C. Cultivation of *O. glaberrima* is believed to be more recent; possibly around 1000 B.C. it was becoming a staple food from the central Niger Delta north to Gambia and Senegal. There is ample evidence that migrating peoples carrying seed with them for cultivation greatly facilitated the broad dispersal of rice as known today.

It is generally agreed that of the three recognized races of *O. sativa*, the Indica and Japonica types are the most important. Indicas, generally from tropical areas, are characterized by tall, vigorous plants with large leaves and a large, firm-cooking grain. Japonicas are usually from temperate areas, are higher yielding, and have short to medium grains that are softer upon

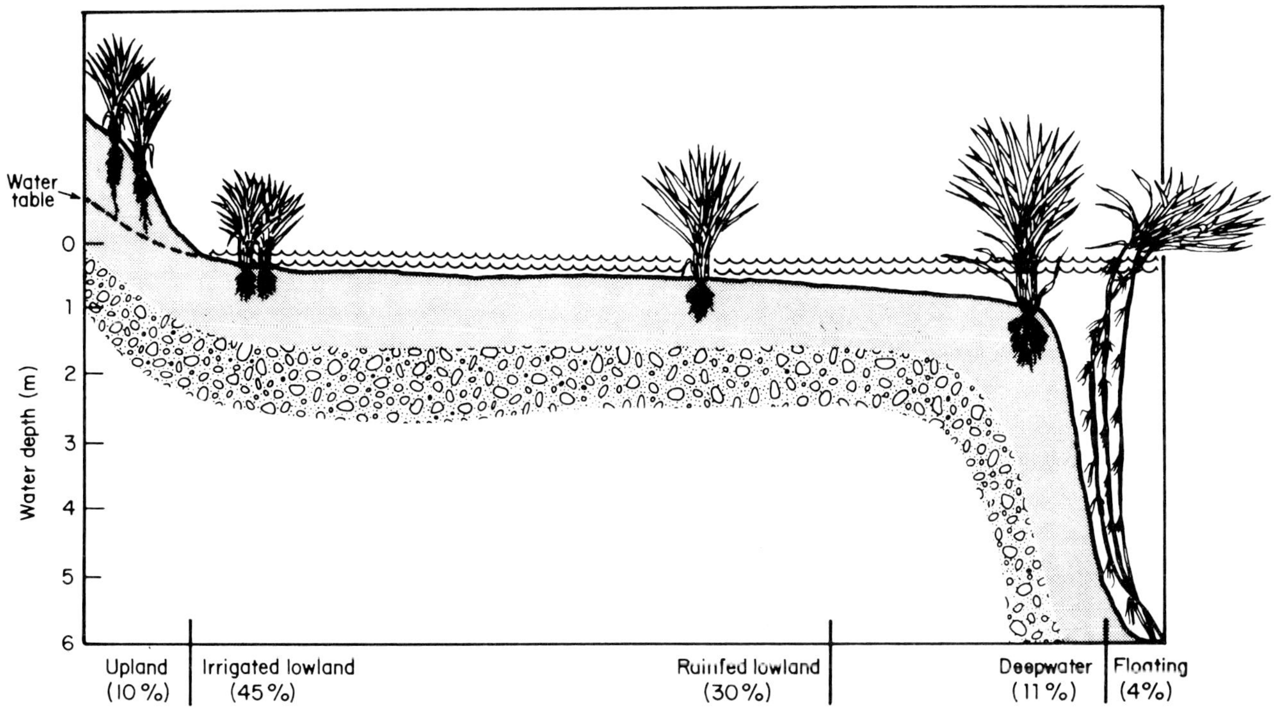

Fig. 1. The world's rice production lands as classified by water regimes used for production and types of rice. (Reprinted, by permission, from DeDatta 1981; copyright © 1981 John Wiley & Sons, Inc.)

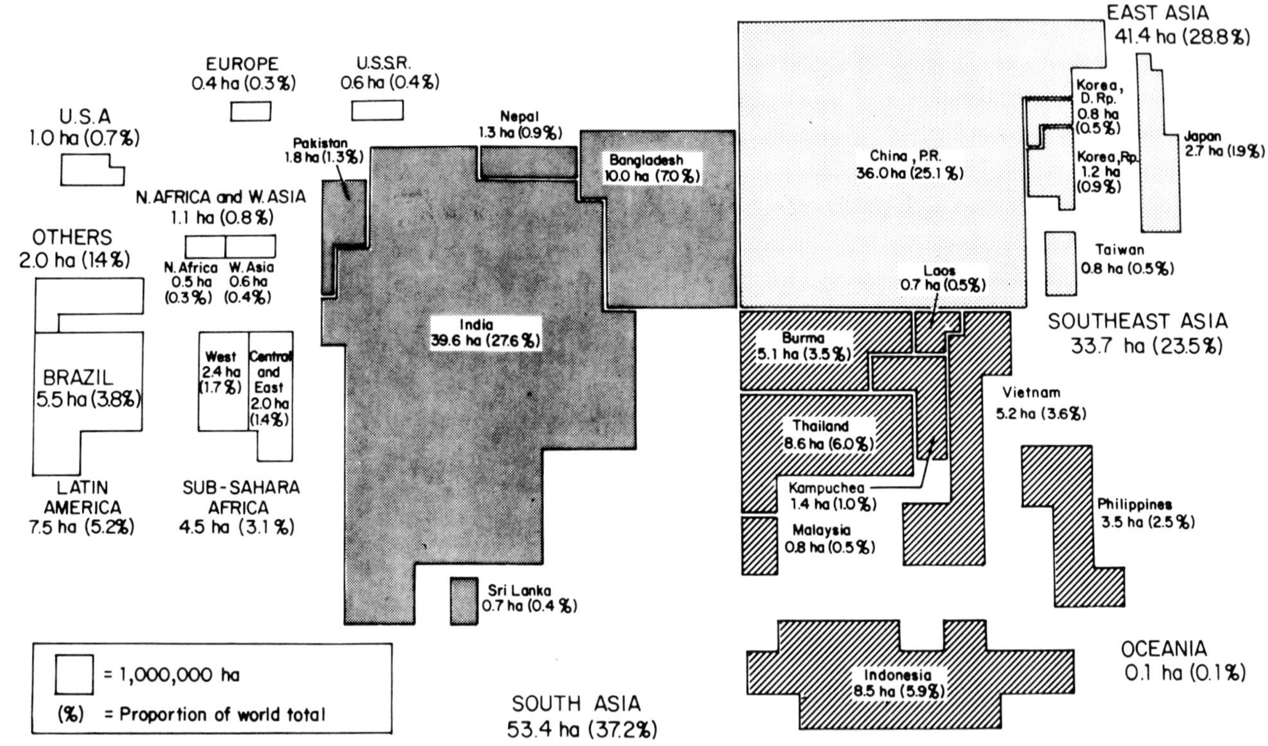

Fig. 2. The world's rice land. (Reprinted, by permission, from DeDatta 1981; copyright © 1981 John Wiley & Sons, Inc.)

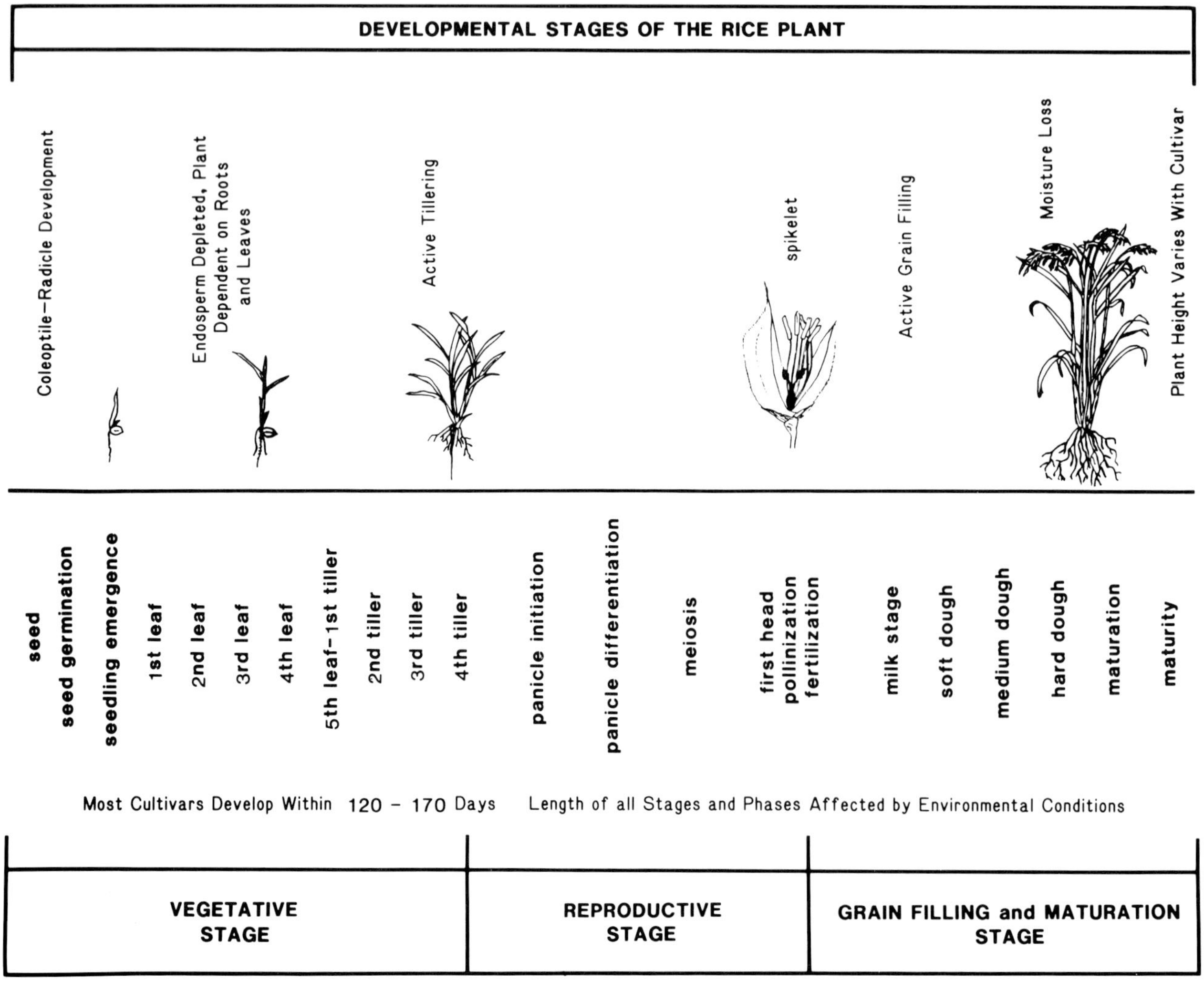

Fig. 3. Developmental stages of the rice plant.

cooking. Selection and improvement through plant breeding and exchange of germ plasm among scientists have, in some cases, led to the development of intermediate types. The extensive efforts in germ plasm collection and breeding have resulted in the accumulation of over 50,000 cultivars and lines in just the collection at the International Rice Research Institute in the Philippines. Thousands of others are held in other collections.

The importance of rice on a global basis cannot be overemphasized. Volumes have been written regarding the role of rice in development of world civilization and societies. Concurrently, an equally large scientific literature has evolved. The comprehensive treatments by S. K. DeDatta and B. S. Luh are recommended to all interested in or working with this intriguing crop.

Selected References

Anonymous. 1990. IRRI Rice Facts. International Rice Research Institute, Los Baños, Laguna, Philippines.

DeDatta, S. K. 1981. Principles and Practices of Rice Production. John Wiley & Sons, New York.

Luh, B. S., ed. 1980. Rice: Production and Utilization. AVI Publishing, Westport, CT.

(Prepared by R. K. Webster)

The Rice Plant

Oryza sativa L. is a diploid grass (24 chromosomes) with round, jointed culms, flat leaves, a fibrous root system, and terminal panicles. Typically, *O. sativa* is an annual plant, but in some tropical areas where climatic conditions are favorable it grows as a perennial. Rice is generally cultivated in aquatic or semiaquatic environments, but it cannot survive without oxygen. Rice is able to thrive in flooded soil because of an internal system of air spaces that allows oxygen to diffuse from the leaf stomata to meristematic tissue in the roots. *O. sativa* also has a capacity for limited anaerobic respiration during certain portions of the growth cycle.

The growth and development of the rice plant is a complex and dynamic process influenced by many factors. Thus, predictions of when various stages of development are expected must account for temperature, seeding dates, cultivar, nutritional status, cultural practices, etc. A generalized diagram depicting the development (with assumptions) of a non-photoperiod-sensitive cultivar is given in Fig. 3. Rice plant development can be divided into three stages: vegetative, reproductive, and grain filling and maturation (ripening). Each of these stages may be further divided into phases for convenience in discussing development and citing morphological parts of the plant (Fig. 4).

Vegetative Stage (Seed Germination to Formation of Panicle Primordia)

The vegetative stage begins with seed germination. The radicle emerges first from the seed embryo, followed by the coleoptile. During the following pretillering phase, seminal and lateral roots and the first leaves develop. These early processes are solely dependent on the endosperm food base. Depending on light conditions, the mesocotyl develops almost concurrently, forming the crown of the plant at the soil surface and giving rise to the main root.

The tillering phase of vegetative growth usually begins when the seedling has developed four or five leaves and is self-supporting. The first tiller generally emerges from an axillary bud in one of the first or lowest nodes of the main stem. Additional tillers are formed (one for each one or two leaves), each displacing a leaf as it develops. These primary tillers give rise to secondary tillers at their nodes, and in turn, secondary tillers may produce tertiary tillers. At maximum tillering, it is often difficult to identify the main stem of the plant. At first, tillers are dependent on the main stem, but after producing leaves and roots they become self-sufficient. Not all of the tillers produced by the plant survive, largely because of competition for space, nutrients, and light.

The first roots produced by the rice plant are usually short-lived, being replaced by secondary crown roots arising from underground nodes on the stem. Superficial adventitious roots may also be produced from nodes above ground as the plant continues to develop.

Reproductive Stage (Panicle Primordia Initiation to Heading)

Panicle initiation generally occurs after the maximum tiller number is reached, signaling the beginning of the reproductive stage. Panicle primordia, or buds, are formed first in the main tillers and are visible as furry tips at the growing point within the culm. Panicle initiation is closely associated with jointing, or internode elongation, which is characterized by the elongation of the uppermost internodes and the concurrent formation of internal air spaces. The timing of this phase and the length of internodes vary by cultivar. Water depth and nutritional status are known to affect this stage of development. It is also during panicle initiation that extensive development of the root system at the soil surface occurs.

As the reproductive stage proceeds, panicle branches, spikelets, and floral parts develop and become distinguishable. The later part of panicle development includes the booting phase, followed by emergence of the panicle (heading) from the flag leaf sheath.

The flowering phase begins soon after the panicles emerge. Flowering progresses from the top to the bottom of the panicles and lasts 6–8 days. The flowers generally open around noon and remain open for only a couple of hours. During this time, anthers on elongated filaments are extended outside the flower, and mature pollen is deposited onto the feathery stigma. Pollen tubes then quickly emerge and enter the stigma. The fertilization process is completed in approximately 2 hr. Flowers are normally self-pollinating (outcrossing occurs in fewer than 1%). The egg is visible within a few days after fertilization.

Grain Filling and Maturation (Ripening)

After fertilization, the rice grain develops through the milk, soft dough, hard dough, mature green, and fully ripe phases. The time from flowering to maturity (ripening) varies with cultivar and location.

As rice grains ripen, leaves begin yellowing due to senescence. As spikelets turn yellow and grain develops to the mature stage, panicles arch with the increased weight of the grain. Harvest maturity is attained as the grain loses moisture and the greenish color fades. The endosperm becomes uniformly hardened and translucent. At this time, the moisture content of grains on main stems usually ranges from 15 to 18%, and total crop grain moisture from 18 to 21%. In some cultivars, the culms remain green during ripening and harvest. When favorable environmental conditions follow harvest, new tillers may form from the stubble, producing ratoon crops.

Selected References

Chang, T., and Bordenas, E. A. 1965. The Morphology and Varietal Characteristics of the Rice Plant. International Rice Research Institute, Los Baños, Laguna, Philippines.

DeDatta, S. K. 1981. Principles and Practices of Rice Production. John Wiley & Sons, New York.

Luh, B. S., ed. 1980. Rice: Production and Utilization. AVI Publishing, Westport, CT. 925 pp.

Stansel, J. W. 1975. The rice plant—Its development and yield. Pages 9-12 in: Six Decades of Rice Research in Texas. Tex. Agric. Exp. Stn. Res. Monogr. 4.

Yoshida, S. 1981. Fundamentals of Rice Crop Science. International Rice Research Institute, Los Baños, Laguna, Philippines.

(Prepared by R. K. Webster)

Diseases of Rice

Diseases of the rice plant are major constraints to production in most areas where the crop is grown. Infectious diseases are those caused by pathogens or transmissible biological agents, such as fungi, bacteria, viruses, mycoplasmas, and nematodes. Noninfectious or abiotic diseases result from unfavorable environmental or nutritional conditions, such as deficiencies or excesses of nutrients and water, temperature extremes, air pollutants, extremes of soil acidity or alkalinity, toxins, and mechanical injury.

The rice plant in all phases of growth is subject to diseases that reduce the quality and quantity of the crop. The extent and severity of disease depends on the presence of a virulent pathogen, a disease-conducive environment, and the suscep-

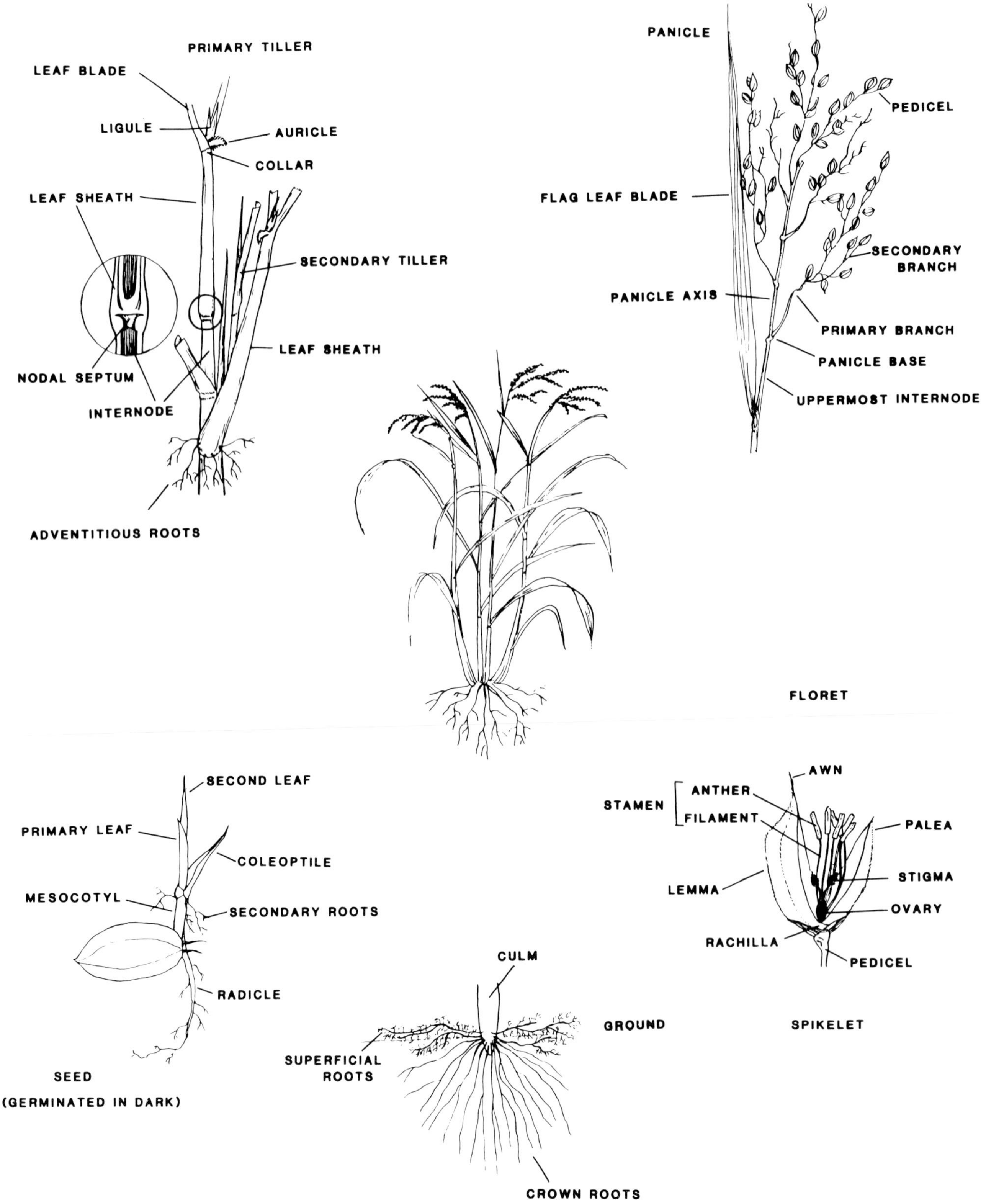

Fig. 4. Morphological parts of the rice plant: seedling, stem, root, inflorescence, and flower.

TABLE 1. Relative Occurrence and Severity of Some of the Most Prevalent Rice Diseases in Major Rice-Growing Regions[a]

Disease	Asia		America		Africa	
	Temperate	Tropical	North	Latin	West	East
Fungal diseases						
Blast	3+++[b]	3+++	2++	3+++	3+++	2++
Sheath blight	3+++	3+++	2++	1++	1+	...[c]
Brown spot	2+++	2+++	2++	1++	1+	1+
Cercospora leaf spot	2+++	2+++	2++	2++	1+	1+
Stem rot	2+++	2+++	2++	1+	1++	...
Bakanae disease	2+++	1++	1+	1+	1+	...
Sheath rot	2++	2++	1+	1+	...	...
Leaf scald	1+	1++	—	1++	1++	1+
Seedling disease	1++	1+	2+	...	...	...
False smut	1+	1+++	1+	1+++	1++	1+
Kernel smut	1+	1++	1++	1++	...	...
Bacterial diseases						
Leaf blight	3+++	3+++	—	1+	1+	2+
Leaf streak	—	2+++	—	—	—	—
Virus diseases						
Dwarf	3++	—	—	—	—	—
Stripe	3++	—	—	—	—	—
Black-streaked dwarf	1	—	—	—	—	—
Yellow dwarf	2+++	1+++	—	—	—	—
Tungro	—	3+++	—	—	—	—
Grassy stunt	—	2+++	—	—	—	—
Transitory yellowing	—	2+	—	—	—	—
Hoja blanca	—	—	±	3+++	—	—
Yellow mottle	—	—	—	—	1+	1+
Nematode diseases						
White tip	2++	2+++	2++	2+++	2++	2++
Stem nematode	—	2++	—	—	—	—
Root nematode	1++	1++	1++	1+	1+	1+

[a] Adapted from Ou, 1980.
[b] Coding for disease severity and occurrence is as follows: 1 = of minor importance, 2 = of moderate importance, 3 = of major importance, — = not present, ± = absent or rarely present, + = occasionally present, ++ = commonly present, +++ = abundantly present.
[c] No data.

tibility of the cultivar. In diseases caused by viruses, mycoplasmas, and some bacteria, the presence of a vector or transmitters of the pathogen is also necessary. This requirement of a vector has resulted in the absence of virus diseases in some areas even though the cultivars grown are susceptible. Likewise, absence of a conducive environment has restricted geographic distribution of some diseases.

The actual number of distinct rice diseases is not clear; over 80 biotic and abiotic diseases have been well described and are included here. A major factor contributing to the uncertainty about the total number of rice diseases stems from the wide distribution of rice in countries where different languages and restricted availability of literature have hampered scientific exchange. This has been particularly exacerbated for fungus diseases; there are many cases of overlap, where different taxonomic names have been given to the same pathogen. The various levels of rigor applied in demonstrating actual pathogenicity (disease caused) have further complicated the situation, as have reports of associations of fungi with rice plants and grain.

Table 1 presents a summary of the general distribution and relative importance of the most common diseases of rice. Although some values given are estimates, they conform generally well with data available in the literature.

Although much commendable work has been done, there are still major challenges ahead in gaining increased understanding of rice diseases and in devising methods to control or minimize their effects on this most significant food crop. Rice diseases are considered individually in this compendium with the hope of aiding the diagnosis and recognition of each, which in turn should facilitate control efforts.

Selected Reference

Ou, S. H. 1980. Rice plant diseases. Pages 235-260 in: Rice Production and Utilization. B. S. Luh, ed. AVI Publishing, Westport, CT.

(Prepared by R. K. Webster)

Part I. Biotic Diseases

Bacterial Diseases

In this section, bacterial diseases of rice are classified into three main groups: seedling, sheath, and grain diseases; foliar diseases; and culm and root disease. The principal diseases in each of these groups are caused by bacteria that belong to a single genus. Diseases of the seedling, sheath, and grain are caused by *Pseudomonas* species; foliar diseases are caused by two *Xanthomonas* species; and a species of *Erwinia* is responsible for a culm and root disease of rice.

Seedling, Sheath, and Grain Diseases

Seedling, leaf sheath, and grain diseases of rice are caused almost exclusively by five species of *Pseudomonas*: *P. plantarii*, *P. avenae*, *P. glumae*, *P. fuscovaginae*, and *P. syringae*. The *Pseudomonas* species are primarily seedborne and cause a necrosis or rot of the plant part affected. With the possible exception of *P. plantarii*, the various species may attack both seedlings and the grain and/or sheaths of mature plants. The diseases caused by the pseudomonads and their geographic distribution are summarized in Table 2. Phenotypic characteristics useful in distinguishing the different species are presented in Tables 3 and 4.

Selected References

Azegami, K., Nishiyama, K., Watanabe, Y., Kadota, I., Ohuchi, A., and Fukazame, C. 1987. *Pseudomonas plantarii* sp. nov., the causal agent of rice seedling blight. Int. J. Syst. Bacteriol. 37:144-152.

Zeigler, R. S., and Alvarez, E. 1990. Characteristics of *Pseudomonas* spp. causing grain discoloration and sheath rot of rice, and associated pseudomonad epiphytes. Plant Dis. 74:917-922.

(Prepared by T. W. Mew)

Seedling Diseases

In this section, seedling blight caused by *Pseudomonas plantarii* and bacterial brown stripe caused by *P. avenae* or *P. syringae* pv. *panici* are covered in depth. The pathogens *P. glumae* and *P. fuscovaginae* and the symptoms they cause on seedlings are treated under sheath and grain diseases.

The incidence and severity of seedling diseases caused by pseudomonads have increased in recent years in areas that use machines for transplanting seedlings. Seedlings for machine transplanting are raised in nursery boxes under conditions of high temperature and humidity that are conducive to the bacterial diseases (Plate 1).

Control of Bacterial Diseases in the Nursery

Chemical treatment of rice seed has been useful in reducing seed infestation and disease caused by *P. glumae*. Kasugamycin used as a seed soak or applied to nursery soil before planting has been the most effective chemical treatment. Cultural practices that are recommended for control of bacterial diseases

TABLE 2. Principal Bacterial Diseases Reported to Affect Rice Seedlings, Sheath, and Grain[a]

Symptom Type	Disease Name	Causal Agent	Distribution
Seedling necrosis	Bacterial brown stripe	*Pseudomonas syringae* pv. *panici*	Asia
	Bacterial brown stripe	*P. avenae*	Worldwide
	Brown stripe	*Erwinia* sp.	Brazil
	Seedling rot	*P. glumae*	Japan, South America
	Seedling blight	*P. plantarii*	Japan
Sheath rot	Bacterial sheath rot	*P. syringae* pv. *syringae*	Hungary, Australia, Asia, Chile
	Bacterial sheath brown rot	*P. fuscovaginae*	Worldwide
Grain discoloration	Black rot	*E. herbicola*	Japan
	Bacterial grain rot	*P. glumae*	Japan, Latin America
	Glume blotch	*P. syringae* pv. *syringae*	Hungary, Australia
	Manchado de grano	*P. fuscovaginae*	Worldwide
	Grain rot	*P. syringae* pv. *aptata*	Japan

[a]Adapted from Zeigler and Alvarez, 1990.

in the nursery include eliminating infected seed by discarding seeds that float in salt water with a specific gravity of 1.18, avoiding overwatering, ensuring a level soil surface and soil with good drainage so that water does not pool, and maintaining the temperature between 15 and 30° C.

P. fuscovaginae can be eradicated from seed by a dry heat treatment at 65° C for 6 days.

Selected References

Goto, K., Zeigler, R. S., and John, V. T. 1988. Progress on rice seedborne and seed contaminant bacteria, viruses, and nematodes. Pages 131-150 in: Rice Seed Health. International Rice Research Institute, Manila, Philippines.

Uyematsu, T., Yoshimura, D., Nishiyama, K., Ibaraki, T., and Fuji, H. 1976. Occurrence of bacterial seedling rot in nursery flat caused by grain rot bacterium *Pseudomonas glumae*. (In Japanese; English abstr.) Ann. Phytopathol. Soc. Jpn. 42:310-312.

(Prepared by T. W. Mew)

Seedling Blight

Seedling blight is a relatively new disease that occurs in Japan and is associated with the production of seedlings in nursery boxes.

Symptoms

Early symptoms are characterized by a basal chlorosis and withering of the second or third leaves. Infected seedlings later become reddish brown and desiccated but do not exhibit a soft rot. With severe infection, root growth is retarded and seedlings easily lodge (Plates 2–4).

Causal Organism

Seedling blight is caused by *Pseudomonas plantarii* Azegami et al. The bacterium is a gram-negative, nonsporeforming, nonencapsulated rod, 0.7–1.0 × 1.4–1.9 μm, with one to three polar flagella. On potato-peptone-glucose agar plates, colonies are round with tan centers and translucent margins, and the bacterium produces a weak, reddish brown pigment. Growth on nutrient agar is poor. The optimum and maximum temperatures for growth of the bacterium are 32–35 and 38° C. The bacterium produces no fluorescent pigment on King's medium B, is oxidase positive and arginine dihydrolase negative, and accumulates poly-β-hydroxybutyrate. *P. plantarii* produces a compound called tropolone, which is responsible for the retardation of root growth and leaf chlorosis observed in infected rice seedlings. Other phenotypic characters of the bacterium are listed in Table 3 (Plate 5).

Disease Cycle and Epidemiology

In Japan, the incidence and severity of the disease have increased with the advent of transplanting by machines. Seedlings for transplanting are raised in nursery boxes under conditions of high temperature and humidity that are conducive to infection.

Control

The application of iron compounds suppresses seedling blight, because the production of tropolone is inhibited in the presence of iron.

Selected References

Azegami, K., Nishiyama, K., and Kato, H. 1987. Growth of *Pseudomonas plantarii*, and production of tropolone and proteins under limited iron deficiency. Ann. Phytopathol. Soc. Jpn. 53:411-412.

Azegami, K., Nishiyama, K., Watanabe, Y., Kadota, I., Ohuchi, A., and Fukazawa, C. 1987. *Pseudomonas plantarii* sp. nov., the causal agent of rice seedling blight. Int. J. Syst. Bacteriol. 37:144-152.

Azegami, K., Nishiyama, K., Watanabe, Y., Suzuki, T., Yoshida, M., Nose, K., and Toda, S. 1985. Tropolone as a root growth-inhibitor produced by a plant pathogenic *Pseudomonas* sp. causing seedling blight of rice. Ann. Phytopathol. Soc. Jpn. 51:315-317.

Zeigler, R. S., and Alvarez, E. 1990. Characteristics of *Pseudomonas* spp. causing grain discoloration and sheath rot of rice, and associated pseudomonad epiphytes. Plant Dis. 74:917-922.

(Prepared by T. W. Mew)

Bacterial Brown Stripe

Bacterial brown stripe, also known as bacterial stripe, occurs in upland and wetland nurseries and also in nursery boxes. Bacterial brown stripe symptoms are caused by both *Pseudomonas avenae* Manns and *P. syringae* pv. *panici* (Elliott) Young et al. *P. avenae* appears to be distributed worldwide, although seedling infection caused by the bacterium has only been reported from a limited number of countries. Seed testing has detected *P. avenae* on seed lots from all major rice-growing countries of the world, and seeds carrying the pathogen produce seedlings with typical symptoms of the disease. Bacterial brown stripe caused by *P. syringae* pv. *panici* has only been reported from China and Japan.

Symptoms

Symptoms consist of water-soaked stripes on the leaves and leaf sheaths, which turn brown. On the leaves, the stripes occur interveinally or along the midrib or leaf margins. The stripes or lesions measure up to 10 cm × 1 mm but may coalesce to form wider lesions. The pathogen can also attack and rot young, unfolded leaves, which may result in stunting or death of the seedling. If only the primary leaves are affected, the seedlings generally outgrow the disease (Plates 6 and 7).

P. avenae can also cause a grain discoloration of mature plants.

Causal Organisms

P. avenae is a gram-negative, nonsporeforming, nonencapsulated rod, 0.92–2.4 × 0.5–0.7 μm, with one or two polar flagella. The bacterium produces no fluorescent pigment on King's medium B and is positive for oxidase, nitrate reduction, and starch hydrolysis.

P. syringae pv. *panici* (syn. *P. panici* (Elliott) Stapp) is also a gram-negative, nonsporeforming, nonencapsulated rod. The bacterium produces a fluorescent pigment on King's medium B and is negative for oxidase, nitrate reduction, and starch hydrolysis.

Disease Cycle and Epidemiology

P. avenae is seedborne, and *P. syringae* pv. *panici* is likely seedborne. Natural infection of *Panicum miliaceum* L., *Hordeum vulgare* L., and *Setaria italica* (L.) P. Beauv. by *P. syringae* pv. *panici* has been reported.

TABLE 3. Phenotypic Characters Useful in Differentiating Nonfluorescent *Pseudomonas* spp. Pathogenic to Rice[a]

	P. plantarii	*P. avenae*	*P. glumae*
Arginine dihydrolase	–	–	–
Oxidase reaction	+	+	–
Nitrate reduction	+	+	–
Starch hydrolysis	–	+	–
Lecithinase	+	–	+
Utilization of:			
L-Arginine	+	–	+
Inositol	+	–	+

[a]Adapted from Azegami et al, 1987, and Zeigler and Alvarez, 1990.

Selected References

Bradbury, J. F. 1985. Guide to Plant Pathogenic Bacteria. CAB International Mycological Institute, Kew, England.
Goto, K., and Ohata, K. I. 1961. Bacterial stripe of rice. Spec. Publ. Coll. Agric. Natl. Taiwan Univ. 10:49-59.
Goto, K., Zeigler, R. S., and John, V. T. 1988. Progress on rice seedborne and seed contaminant bacteria, viruses, and nematodes. Pages 131-150 in: Rice Seed Health. International Rice Research Institute, Manila, Philippines.
Kadota, I., and Ohuchi, A. 1983. Symptoms of bacterial brown stripe of rice seedlings in nursery boxes. Ann. Phytopathol. Soc. Jpn. 49:561-564.
Shakya, D., Vinther, F., and Mathur, S. 1985. World wide distribution of a bacterial stripe pathogen of rice identified as *Pseudomonas avenae*. Phytopathol. Z. 114:256-259.
Tominaga, T. 1968. Brown stripe of bromegrass and wheat grasses caused by *Pseudomonas setariae* (Okabe) Savulescu. Jpn. J. Bacteriol. 23:176-183.
Young, J. M. 1988. Comment on the proposal of Hildebrand & Palleroni (1987) to reject the name *Pseudomonas syringae* pv. *panici* (Elliott 1923) Young et al. 1978. Lett. Appl. Microbiol. 88:47-50.
Zeigler, R. S., and Alvarez, E. 1990. Characteristics of *Pseudomonas* spp. causing grain discoloration and sheath rot of rice, and associated pseudomonad epiphytes. Plant Dis. 74:917-922.

(Prepared by T. W. Mew)

Leaf Sheath and Grain Diseases

Pseudomonas fuscovaginae, *P. syringae* pv. *syringae*, *P. glumae*, and *P. avenae* attack the flag leaf sheaths and grain of rice and are distributed worldwide. *P. fuscovaginae* and *P. syringae* pv. *syringae* typically cause a severe sheath rot and/or extensive spikelet sterility and grain discoloration. *P. glumae* and *P. avenae* are primarily associated with spikelet sterility and grain discoloration but may also cause mild sheath rot symptoms. The symptoms produced by the four species on rice grain are similar.

Other reports of pathogens that cause grain discoloration of rice include *P. syringae* pv. *aptata* (Brown & Jamieson) Young et al, *Erwinia amylovora* (Burrill) Winslow et al, *E. carotovora* (Jones) Bergey et al, *Xanthomonas cinnamona* (Miyake & Tsunoda) Muko, and *X. atroviridigerum* (Miyake & Tsunoda) Tagami & Mizukami. Little information on the diseases caused by these pathogens is available, and the identity of the pathogens and their pathogenicity is questionable.

Sheath Brown Rot

Sheath brown rot, caused by *Pseudomonas fuscovaginae* Tanii, Miyajima, & Akita, has been reported in Asia, Latin America, South America, Central Africa, and Madagascar. The disease is manifested as a sheath and grain rot of mature plants as well as a seedling rot and has caused substantial yield losses in South America.

TABLE 4. Characters Useful in Differentiating Fluorescent *Pseudomonas* spp. Associated with Sheath Rot and Grain Discoloration of Rice[a]

	P. fuscovaginae	*P. syringae*	*P. fluorescens*
Arginine dihydrolase	+	−	+
Oxidase reaction	+	−	+
Nitrate reduction	−	−	+
Levan production from sucrose	−	+	−
Utilization of inositol	−	+	−
Tobacco hypersensitive reaction	+	+	−

[a]Adapted from Zeigler and Alvarez, 1990.

Symptoms

On seedlings, a systemic discoloration of the leaf sheath occurs, which may spread to the midrib or veins of the leaves. On mature plants, symptoms typically occur on the flag leaf sheath from the booting to heading stage and on the panicle (Plate 8). On the sheath, oblong to irregular dark green, water-soaked lesions occur, which become gray-brown or brown and may be surrounded by an effuse dark brown margin (Plate 9). The sheath may also exhibit general water-soaking and necrosis without definable lesions. With severe infections, the entire leaf sheath may become necrotic and dry out, and the panicle withers. Glumes of panicles emerging from infected sheaths exhibit water-soaked lesions that turn light brown. Grains of infected panicles are discolored, deformed, or empty.

Causal Organism

The causal bacterium of sheath brown rot was originally described as *P. marginalis* (*P. fluorescens* biovar II). Later, the pathogen was described as a new species, *P. fuscovaginae* Tanii, Miyajima, & Akita. Strains of *P. fuscovaginae* are reportedly heterogenous and fall into two groups. Whether or not these groups represent distinct species has not been determined. *P. fluorescens* is not pathogenic to rice and is clearly different from *P. fuscovaginae.*

P. fuscovaginae is a gram-negative, nonsporeforming rod, 0.5–0.8 × 2.0–3.5 μm, with one to four polar flagella. After 4–5 days at 28° C, colonies on nutrient agar are white to cream-colored, smooth, convex, translucent, butyrous, and 3–5 mm in diameter. The bacterium produces a fluorescent pigment on King's medium B and is positive for oxidase and arginine dihydrolase.

Phenotypic characters that distinguish the different fluorescent pseudomonads associated with sheath rot are presented in Table 4.

Disease Cycle and Epidemiology

P. fuscovaginae survives on rice seed at a low level and as an epiphyte on gramineous weeds in rice-growing areas. In inoculation studies, the bacterium has been found to be pathogenic to wheat, oats, barley, triticale, corn, *Lolium perenne* L., *Bromus marginatus* Nees ex Steudel, *Phleum pratense* L., and *Phalaris arundinacea* L., and these hosts may serve as reservoirs of inoculum in the field.

The bacterium can be recovered from healthy rice leaf blades and sheaths in the field, and epiphytic populations of the bacterium peak at the booting stage. Disease development on mature plants is favored by cool daytime temperatures (17–23° C) that delay panicle emergence. The seedling rot phase of the disease is most prevalent at temperatures below 20° C.

Control

Use of clean seed or seed treated with dry heat at 65° C for 6 days is important in controlling sheath brown rot. Also, antibiotics such as streptomycin, alone or in combination with oxytetracycline, can effectively control sheath brown rot if applied at or a few days after panicle emergence.

Selected References

Miyajima, K., Tanii, A., and Akita, T. 1983. *Pseudomonas fuscovaginae* sp. nov., nom. rev. Int. J. Syst. Bacteriol. 33:656-657.
Tanii, A., Miyajima, K., and Akita, T. 1976. The sheath brown rot disease of rice plant and its causal bacterium, *Pseudomonas fuscovaginae*, A. Tanii, K. Miyajima et T. Akita sp. nov. Ann. Phytopathol. Soc. Jpn. 42:540-548.
Zeigler, R. S., and Alvarez, E. 1987. Bacterial sheath brown rot of rice caused by *Pseudomonas fuscovaginae* in Latin America. Plant Dis. 71:592-597.
Zeigler, R. S., and Alvarez, E. 1990. Characteristics of *Pseudomonas* spp. causing grain discoloration and sheath rot of rice, and associated

pseudomonad epiphytes. Plant Dis. 74:917-922.
Zeigler, R. S., Aricapa, G., and Hoyos, E. 1987. Distribution of fluorescent *Pseudomonas* spp. causing grain and sheath discoloration of rice in Latin America. Plant Dis. 71:896-900.

(Prepared by T. W. Mew)

Sheath Rot

Sheath rot, caused by *Pseudomonas syringae* pv. *syringae* van Hall (syn. *P. oryzicola* Klement), is identical in symptomatology to sheath brown rot, caused by *P. fuscovaginae* Tanii, Miyajima & Akita. It is the only reported sheath rot pathogen of rice in Chile and has also been reported from Asia, Australia, and Hungary.

P. syringae produces a fluorescent pigment on King's medium B but is easily differentiated from *P. fuscovaginae* because it is negative for oxidase and arginine dihydrolase.

Selected Reference

Zeigler, R. S., and Alvarez, E. 1990. Characteristics of *Pseudomonas* spp. causing grain discoloration and sheath rot of rice, and associated pseudomonad epiphytes. Plant Dis. 74:917-922.

(Prepared by T. W. Mew)

Grain Rot

Grain rot occurs in Japan, Korea, and Taiwan. The disease is manifested as a grain rot of mature plants in the field and also as a seedling rot.

Symptoms

On seedlings, symptoms consist of a brown, water-soaked soft rot of the leaf sheaths accompanied by wilting or soft rot of the leaves (Plate 10). On the panicle, infected grains are shrunken and pale green, becoming dirty yellow to brown and dry (Plate 11). The infected portions of the lemma and palea are purplish brown to dark brown. In light infections, only the palea is discolored. A brown margin between the infected and healthy parts of the grain is a diagnostic feature of the disease. No symptoms are present on the panicle rachis and branches. The bacterium may cause a mild rot of the flag leaf sheath or the flag leaf sheath collar.

Causal Organism

Grain rot is caused by *Pseudomonas glumae* Kurita & Tabei. The bacterium is a gram-negative rod, 0.5–0.7 $\times$ 1.5–2.5 μm, with one to three polar flagella. It forms white colonies on nutrient agar and produces no fluorescent pigment on King's medium B. Other phenotypic characters of the bacterium are listed in Table 3.

A selective medium, S-PG, useful for isolation of the bacterium is KH_2PO_4 (1.3 g), Na_2HPO_4 (1.2 g), $(NH_4)_2SO_4$ (5.0 g), $MgSO_4 \cdot 7H_2O$ (0.25 g), $Na_2MgO_4 \cdot 2H_2O$ (24 mg), Fe-EDTA (10 mg), L-cystine (10 μg), D-sorbitol (10 g), potassium phenethicillin (50 mg), ampicillin sodium (10 mg), cetrimide (10 mg), methyl violet (1 mg), phenol red (20 mg), agar (15 g), and distilled water 1 L. On this agar, *P. glumae* produces circular, convex colonies with smooth margins that are reddish brown or opalescent with a purple or red-purple center.

Disease Cycle and Epidemiology

The bacterium is seedborne and invades the spaces between the cells in the outer epidermis and spongy parenchyma of the lemma. From infected seed it may grow epiphytically at a low level until panicle emergence, when its population increases rapidly on the grain. The disease is favored by high night temperatures (28° C) and high humidity.

Selected References

Chien, C. C., Chang, Y. C., Lao, Y. M., and Ou, S. H. 1983. Bacterial grain rot of rice—A new disease in Taiwan. J. Agric. Res. China 32:360-366.
Tsushima, S., Wakimoto, S., and Mogi, S. 1986. Selective medium for detecting *Pseudomonas glumae* Kurita and Tabei, the causal bacterium of grain rot of rice. Ann. Phytopathol. Soc. Jpn. 52:253-259.
Zeigler, R. S., and Alvarez, E. 1990. Characteristics of *Pseudomonas* spp. causing grain discoloration and sheath rot of rice, and associated pseudomonad epiphytes. Plant Dis. 74:917-922.

(Prepared by T. W. Mew)

Bacterial Palea Browning

Bacterial palea browning occurs in various districts in Japan and affects grain quality. Disease incidences as high as 32% and reductions in 1,000-seed weight by as much as 15% have been reported in some localities.

Symptoms

Symptoms usually first appear at early flowering. Initially, light brown, water-soaked lesions occur on the lemma or palea (Plate 12A). The lesions then turn dark brown. The discoloration occurs most frequently on the palea (Plate 12B). Infected panicles have more immature grains and lighter grains at harvest, and infected grain becomes brown after milling.

Causal Organism

The bacterium responsible for palea browning has been identified as *Erwinia herbicola* (Löhnis) Dye, although it has affinities with *E. ananas* Serrano. The bacterium is gram-negative and fermentative, with peritrichous flagella, and produces a yellow water-soluble pigment. It resembles *E. ananas* in that it produces acid from melibiose, cellobiose, and glycerol, but not from dextrin; produces indole; does not reduce nitrate; and is phenylalanine deaminase negative.

Disease Cycle and Epidemiology

High epiphytic populations of *E. herbicola* are common on rice. The disease occurs when heading coincides with periods of rain and high temperatures in the range of 30–35° C. Disease incidence is increased in fields with high levels of nitrogen fertilization, especially at heading.

Control

No control for the disease is available.

Selected Reference

Azegami, K., Ozaki, K., Matsuda, A., and Ohata, K. 1983. Bacterial palea browning, a new disease of rice caused by *Erwinia herbicola*. (In Japanese; English abstr.) Bull. Natl. Inst. Agric. Sci. Ser. C 37:1-12.

(Prepared by T. W. Mew)

Foliar Diseases

Bacterial Blight

Bacterial blight is one of the most serious diseases of rice worldwide and is found in both tropical and temperate regions. It was first recorded in Japan in 1884. In the 1960s, bacterial blight became prevalent in other rice-growing regions of Asia with the introduction of "improved" cultivars such as TN1 and IR8, which were susceptible to the disease. In addition to Asia, bacterial blight occurs in Australia, Africa, Latin America, the Caribbean, and the United States. Economically, it has had the greatest impact in Asia, where several epidemics have occurred in the past two decades, and in West Africa, particularly in Niger, where irrigated rice was extensively damaged in 1982. Yield losses of 10–50% from leaf blight have been reported.

Symptoms

There are two main syndromes caused by bacterial blight—leaf blight and wilt, or kresek. Leaf blight, the more common syndrome, generally occurs from the maximum tillering stage onward (Plate 13). It begins as water-soaked stripes on the leaf blades (Plate 14A). The stripes increase in length and width, become yellow and then white, and may coalesce to cover the entire leaf blade (Plate 14B). Drops of bacterial exudate may be observed on young lesions. Older infected leaves may appear grayish from the growth of saprophytic fungi. Small, circular lesions with water-soaked margins may also form on the glumes with severe infections. Infected plants produce fewer and lighter grains, and the grain is of poor quality.

The wilt syndrome, known as kresek, is the most destructive manifestation of the disease; it occurs in the tropics from the seedling to the early tillering stage (Plate 15). Leaves of infected plants wilt and roll up, turning grayish green. The leaves then turn yellow to straw-colored and wither, and the entire plant generally dies. Plants that do survive are stunted and yellowish. Total crop failure is not uncommon with kresek.

A third syndrome associated with bacterial blight is called yellow leaf or pale yellow (Plates 16 and 17). The youngest leaf of the plant becomes uniformly pale yellow or has a broad yellow stripe. With yellow leaf, the bacteria are not present in the leaf itself but can be found in the internodes and crowns of affected stems.

Causal Organism

Bacterial blight is caused by *Xanthomonas oryzae* pv. *oryzae* (ex Ishiyama) Swings et al (syn. *X. campestris* pv. *oryzae* (Ishiyama) Dye). The bacterium is a gram-negative, nonspore-forming rod, 0.55–0.75 × 1.35–2.17 μm. On nutrient agar, the colonies are light yellow, circular, convex, and smooth, and they produce a yellow water-soluble pigment. Phenotypic characters useful in distinguishing *X. oryzae* pv. *oryzae* from *X. oryzae* pv. *oryzicola* (Fang et al) Swings et al, which causes bacterial leaf streak of rice and can mimic bacterial blight, are listed in Table 5. The bacterium is systemic in the xylem of the rice host.

Pathogenic Variation. *X. campestris* pv. *oryzae* is variable in virulence, and several races of the bacterium have been described, primarily in Japan and the Philippines. At this time, there is no international set of differentials suitable for race identification in all countries. Race identification is also confounded by problems with inoculation methodologies and the fact that disease responses are not always consistent. Strains of *X. campestris* pv. *oryzae* from Texas and Louisiana form a low-virulence group compared with Philippine strains on the host Taichung Native 1.

A monoclonal antibody specific to *X. campestris* pv. *oryzae* and a second, specific only to strains from Texas and Louisiana, have been developed. To date, monoclonal antibodies specific to races of the bacterium have not been produced, but race-specific restriction length polymorphisms have been reported.

Disease Cycle and Epidemiology

X. oryzae pv. *oryzae* survives primarily in rice stubble and on weed hosts, notably *Leersia oryzoides* (L.) Sw., *Zizania latifolia* (Griesb.) Turcz. ex Stapf., *Leptochloa chinensis* (L.) Nees, *L. panicea* (Retz.) Ohwi, and *Cyperus rotundis* L. In Australia, the bacterium is known to survive on the wild *Oryza* species *O. rufipogon* Griff. and *O. australiensis* Domin. It can also survive for short periods on infected seed and in soil, but these have not been demonstrated to be important sources of inoculum. In tropical areas, it may survive in irrigation water.

The bacterium infects the host through hydathodes or wounds in the leaves or roots and multiplies in the xylem. The kresek syndrome is generally associated with seedling infection that occurs through wounds made during transplanting operations. The roots of seedlings are often damaged when they are removed from the nurseries, and in many areas, the tops of the leaves are clipped off before transplanting. The severity of kresek is dependent on the time of infection; the earlier seedlings are infected, the more severe is the syndrome.

Bacterial blight is favored by warm temperatures (25–30° C), high humidity, rain, and deep water and is more prevalent in wetland areas where these conditions often occur. Severe winds, which cause wounds, and excess nitrogen also favor the disease. The severity of the disease is in part dependent on the virulence of the isolates present. The bacterium can be disseminated by irrigation water, by splashing or windblown rain, by plant-to-plant contact, by trimming tools used in transplanting, and by handling during transplanting.

Control

The disease can be effectively controlled with resistant cultivars. Proper breeding strategies and deployment of resistant cultivars are dependent on monitoring the pathogenic specialization of the population present in a given area.

Cultural control recommendations for the disease include avoiding excessive nitrogen fertilization, maintaining shallow water in nursery beds, providing good drainage during severe flooding, plowing under rice stubble and straw following harvest, and removing alternate hosts.

Control of the disease with copper compounds, antibiotics, and other chemicals has not proven effective.

Selected References

Benedict, A. A., Alvarez, A. M., Berestecky, J., Imanaka, W., Mizumoto, C. Y., Pollard, L. W., Mew, T. W., and Gonzalez, C. F. 1989. Pathovar-specific monoclonal antibodies for *Xanthomonas campestris* pv. *oryzae* and for *Xanthomonas campestris* pv. *oryzicola*. Phytopathology 79:322-328.

Jones, R. K., Barnes, L. W., Gonzalez, C. F., Leach, J. E., Alvarez, A. M., and Benedict, A. A. 1989. Identification of low-virulence strains of *X. campestris* pv. *oryzae* from rice in the United States. Phytopathology 79:984-990.

Mew, T. W. 1987. Current status and future prospects of research on bacterial blight of rice. Annu. Rev. Phytopathol. 25:359-382.

Mew, T. W. 1989. An overview of the world bacterial blight situation. Pages 7-12 in: Proc. Bacterial Blight of Rice. International Rice Research Institute, Los Baños, Philippines.

Swings, J., Van den Mooter, M., Vauterin, L., Hoste, B., Gillis, M., Mew, T. W., and Kersters, K. 1990. Reclassification of the causal agents of bacterial blight (*Xanthomonas campestris* pv. *oryzae*) and bacterial leaf streak (*Xanthomonas campestris* pv. *oryzicola*) of rice as pathovars of *Xanthomonas oryzae* (ex Ishiyama 1922) sp. nov., nom. rev. Int. J. Syst. Bacteriol. 40:309-311.

Vera Cruz, C. M., Gossele, F., Kersters, K., Segers, P., Van den Mooter, M., Swings, J., and De Ley, J. 1984. Differentiation between *Xanthomonas campestris* pv. *oryzae*, *Xanthomonas campestris* pv. *oryzicola* and the bacterial "brown blotch" pathogen on rice by numerical analysis of phenotypic features and protein gel electrophoregrams. J. Gen. Microbiol. 130:2983-2999.

(Prepared by T. W. Mew)

Bacterial Leaf Streak

Bacterial leaf streak is widely distributed in tropical Asia and in West Africa in both lowland and upland rice-growing areas. Estimates of yield loss from the disease range from 1 to 17% depending on the cultivar and the climatic conditions. Recently, the disease has been reported in epidemic proportions in southern China.

Symptoms

The disease can occur at any growth stage and initially appears as small, interveinal, water-soaked streaks (Plate 18). The streaks are at first dark green and later become translucent (Plate 19). The streaks enlarge and coalesce and eventually become light brown. Numerous tiny yellow beads of bacterial exudate are common on the surface of lesions (Plate 20). Eventually, entire leaves turn brown and then grayish white and die. The latter stage of the disease can mimic bacterial blight.

Causal Organism

Bacterial leaf streak is caused by *Xanthomonas oryzae* pv. *oryzicola* (Fang et al) Swings et al (syn. *X. campestris* pv. *oryzicola* (Fang et al) Dye). The bacterium is a gram-negative, nonsporeforming rod, 1.2 × 0.3–0.5 μm, with a single polar flagellum. Colonies on nutrient agar are pale yellow, circular, smooth, and convex. Phenotypic characters of the bacterium useful in differentiating it from *X. oryzae* pv. *oryzae*, causal agent of bacterial blight, are listed in Table 5. A monoclonal antibody specific for *X. oryzae* pv. *oryzicola* has been developed. Isolates of the bacterium that differ in virulence have been identified. In the rice host, the bacterium is present in the parenchyma tissue.

TABLE 5. Phenotypic Characters Useful in Differentiating *Xanthomonas oryzae* pv. *oryzae* from *X. oryzae* pv. *oryzicola*

	X. oryzae pv. *oryzae*	*X. oryzae* pv. *oryzicola*
Acetoin production	−	+
Strong peptonization of litmus milk	−	+
Phenylalanine deaminase	−	+[a]
Utilization of L-alanine	−	+
Sensitivity to 0.001% cupric nitrate (w/v)	+	−

[a]Positive response in 50% of strains.

Disease Cycle and Epidemiology

The bacterium survives largely on infested seed and straw. All wild species of the genus *Oryza* can be infected by *X. oryzae* pv. *oryzicola* and may serve as reservoirs of inoculum. No other hosts of the bacterium are known. The bacterium may also be able to survive in irrigation water.

The bacterium enters the host through stomates or wounds and multiplies in parenchymatous tissue. Bacterial exudates from leaf lesions are disseminated primarily by splashing and windblown rain and also by leaf contact and irrigation water. Disease development is favored by rain, high humidity, and high temperatures (28–30° C).

Control

The disease can be controlled with the use of resistant cultivars and treated seed. In practice, growers have relied on resistant cultivars for disease control.

Selected References

Benedict, A. A., Alvarez, A. M., Berestecky, J., Imanaka, W., Mizumoto, C. Y., Pollard, L. W., Mew, T. W., and Gonzalez, C. F. 1989. Pathovar-specific monoclonal antibodies for *Xanthomonas campestris* pv. *oryzae* and for *Xanthomonas campestris* pv. *oryzicola*. Phytopathology 79:322-328.

Ou, S. H. 1985. Rice Diseases, 2nd ed. Commonwealth Mycological Institute, Kew, England.

Swings, J., Van den Mooter, M., Vauterin, L., Hoste, B., Gillis, M., Mew, T. W., and Kersters, K. 1990. Reclassification of the causal agents of bacterial blight (*Xanthomonas campestris* pv. *oryzae*) and bacterial leaf streak (*Xanthomonas campestris* pv. *oryzicola*) of rice as pathovars of *Xanthomonas oryzae* (ex Ishiyama 1922) sp. nov., nom. rev. Int. J. Syst. Bacteriol. 40:309-311.

Vera Cruz, C. M., Gossele, F., Kersters, K., Segers, P., Van den Mooter, M., Swings, J., and De Ley, J. 1984. Differentiation between *Xanthomonas campestris* pv. *oryzicola* and the "brown blotch" pathogen on rice by numerical analysis of phenotypic features and protein gel electrophoregrams. J. Gen. Microbiol. 130:2983-2999.

(Prepared by T. W. Mew)

Halo Blight

Halo blight, caused by *Pseudomonas syringae* pv. *oryzae* Kuwata, was first reported in 1985 and is currently limited to the Aomori Prefecture, Japan. The disease is characterized by circular, pale green to yellowish brown lesions, 2–10 mm in diameter, on leaf blades (Plate 21). The lesions are surrounded by a distinct halo and have a dark brown spot or stripe in the center. The lesions may coalesce to form large blotches. The disease has not caused any serious damage to date.

Selected Reference

Kuwata, H. 1985. *Pseudomonas syringae* pv. *oryzae* pv. nov., causal agent of bacterial halo blight of rice. Ann. Phytopathol. Soc. Jpn. 33:111.

(Prepared by T. W. Mew)

Culm and Root Diseases

Foot Rot

Foot rot occurs in Japan, India, Bangladesh, Korea, and the Philippines.

Symptoms

Leaf sheaths of infected plants typically exhibit a dark brown decay, and the attached leaves turn yellow and wilt. Infection often begins in the ligules. The nodes, culms, and crowns also become rotted, and infected tillers can easily be detached from the crown. Infected culms and internodes turn black (Plate 22). Roots attached to infected nodes decay and fall off. Bacterial ooze may be present inside the culms (Plate 23), and infected plants have an unpleasant odor. In some cases, the young leaves of tillers that show no sheath browning may wilt as a result of systemic infection of the crown alone (Plate 24).

Causal Organism

Foot rot is caused by *Erwinia chrysanthemi* Burkholder, McFadden, & Dimock. The bacterium is a gram-negative rod, with four to six peritrichous flagella. On yeast extract–dextrose–calcium carbonate agar, the bacterium produces brown to dark brown colonies. The optimum and maximum temperatures for growth are 36 and 43° C, respectively. *E. chrysanthemi* isolates from rice are biochemically similar to isolates from corn. However, isolates from corn are weakly virulent on rice, whereas isolates from rice are virulent on corn.

Disease Cycle and Epidemiology

E. chrysanthemi has a wide host range. In Japan, iris plants, which are common around rice fields, were shown to be susceptible to the bacterium and capable of serving as an inoculum source. The bacterium can also be isolated from the rhizosphere of healthy rice plants. The bacterium is thought to be disseminated primarily by irrigation water. The bacterium enters rice plants through wounds in the roots.

Selected References

Goto, M. 1979. Dissemination of *Erwinia chrysanthemi*, the causal organism of bacterial foot rot of rice. Plant Dis. Rep. 63:100-103.

Goto, M. 1979. Bacterial foot rot of rice caused by a strain of *Erwinia chrysanthemi*. Phytopathology 69:213-216.

(Prepared by T. W. Mew)

Fungal Diseases

Fungi have been reported to cause more diseases of rice than any other group of pathogens. In this section, diseases caused by fungi are grouped into seedling, foliar, leaf sheath and culm, root and crown, and grain diseases. The most important of these, such as blast, sheath blight, stem rot, and seedling diseases, can devastate a crop when environmental and cultural conditions favor disease development.

A relatively large number of fungi have been reported to cause minor diseases. These include diseases about which little is known or which are only partially documented and those in which the correct nomenclature of the purported pathogen is unclear. A list of fungi named in such reports is included.

Seedling Diseases

Seedling diseases can affect rice in water-seeded fields, in seedbeds grown for transplanting, and in seed boxes for mechanical transplanting. These diseases cause stand reduction and stand irregularities in directly seeded rice and loss of healthy seedlings for transplanting in seedbeds and seedling boxes. Also, infected seedlings may serve as sources of inoculum for the spread of certain pathogens into older rice, where they may cause serious foliage or panicle diseases. Seed-rotting and seedling diseases can often be ameliorated by the use of pesticides or by cultural practices.

Water Mold

Water mold, also known as seed rot or seedling damping-off, probably occurs wherever rice is directly seeded into water, either in the field or in seedbeds. Most of the research on this disease has been conducted in Japan and the United States. Although water mold has been observed in the tropics, it is primarily a problem where seeds are sown directly into cold water.

Symptoms

In water-seeded rice, water mold can be observed through clear water as a ball of fungal hyphae surrounding seeds on the soil surface (Plate 25). In areas where the fields are drained after sowing, affected seed is typically surrounded by a mass of fungal strands radiating out over the soil surface (Plate 26). A circular copper brown or dark green spot about 25–35 mm in diameter eventually develops around the rotted seed or stunted seedling. The color is caused by bacteria and green algae that are mixed with the fungal hyphae. Hyphal strands are readily visible without magnification.

When seeds are attacked immediately after sowing, they do not germinate, because the endosperm is destroyed or the embryo is rotted. The endosperm becomes liquefied and oozes out as a thick, white liquid when the seed is mashed, and the embryo turns yellow-brown and finally dark brown. When seeds are attacked following germination, the growth of seedlings can be severely affected, with stunting, yellowing, or rotting of young seedlings. If the weather is favorable for plant growth, surviving seedlings often outgrow the disease and are not severely damaged.

In seedbeds that are kept continuously damp but not flooded, the surface of infected seeds has a cottony hyphal growth. The fungi are present in cracks in the seed coat or where the radicle and plumules are emerging from the seed. In general, the same fungi that cause disease in water-seeded fields infect seeds grown in beds for transplanting.

Causal Organisms

In the United States, this disease is caused by a complex of *Achlya* and *Pythium* and at least one *Fusarium* sp. The main species involved are *A. conspicua* Coker, *A. klebsiana* Pieters, *P. dissotocum* Drechs., *P. spinosum* Sawada, and a *Fusarium* sp. In Japan, in addition to species of *Achlya* and *Pythium*, species of *Saprolegnia, Dictyuchus, Pythiogeton*, and *Phytophthora* have been reported to cause water mold. These fungi are waterborne and soilborne.

Disease Cycle and Epidemiology

Between crops, the fungi survive in field soil. The fungi produce zoospores in response to flooding or wetting of the soil. The zoospores are attracted to cracks in the seed coat where the endosperm is exposed or to the radicle and plumules of the germinating seeds. *Achlya* spp. normally attack the endosperm of germinating seeds, whereas *Pythium* spp. usually attack the developing embryo. Water temperatures below 13° C or above 27° C favor disease development. When daily high and low air or water temperatures average above 18° C, seedling growth is favored and water mold is less severe. Seeds with poor vigor and slow growth will usually be damaged by water mold fungi when water-seeded.

Damage by larvae of the midge *Chironomus plumosus* L. in water-seeded rice may be confused with water mold damage and can contribute to increased severity of the disease. Midge larvae damage germinating rice seeds by feeding on the embryo or the developing roots and leaves of very young seedlings.

Control

Seed treatments with a recommended fungicide are effective in controlling water mold. In the United States, most seed sold commercially is already treated. In field tests, stands produced by fungicide-treated seeds were increased by 25% to more than 100% over stands produced by untreated seeds.

Proper water management can also reduce water mold. Water-seeded rice sown into a deep, continuous flood is most susceptible to water mold, because deep water slows seedling development. Draining a field 3 or 4 days after seeding and then flushing (flooding and draining) seedlings as needed reduces stand losses to water mold compared with maintaining a deep, continuous flood. In some areas, however, a continuous flood is necessary for control of certain weed species. In such areas, laser leveling of the soil has enabled growers to maintain a permanent shallow flood that inhibits weeds and minimizes water mold by allowing for rapid seedling development. An alternative to continuous shallow flooding that also minimizes water mold and is also dependent on precisely leveled soil is pinpoint flooding. This method involves seeding into water, draining the field for 3 or 4 days to allow seedlings to be established, and then shallowly flooding the field, gradually increasing the depth as the seedlings grow.

Selected References

Anonymous. 1987. Rice production handbook. La. State Univ. Agric. Cent. Bull. 2321. pp. 36-46.

Ou, S. H. 1985. Rice Diseases. 2nd ed. Commonwealth Mycological Institute, Kew, England.

Rush, M. C. 1985. Controlling seed-rot and seedling diseases in rice. (Abstr.) Phytopathology 75:1329.

Rush, M. C., and Gifford, J. R. 1972. Enhancement of fungicidal control of seed rot and seedling diseases of rice by an insecticidal seed treatment. Plant Dis. Rep. 56:154-157.

Rush, M. C., Marchetti, M. A., and Adair, C. R. 1972. Stand establishment in early-seeded rice in the South. Rice J. 75:32-34.

Rush, M. C., and Schneider, R. H. 1990. Chemical control of seedling diseases of rice in Louisiana. Pages 53-70 in: Pest Management in Rice. B. T. Grayson, M. B. Green, and L. G. Copping, eds. Elsevier Applied Science, New York.

Webster, R. K., Hall, D. H., Heeres, J., Wick, C. M., and Brandon, D. M. 1970. *Achlya klebsiana* and *Pythium* species as primary causes of seed rot and seedling diseases of rice in California. Phytopathology 60:964-968.

Webster, R. K., Hall, D. H., Wick, C. M., and Brandon, D. M. 1973. Chemical seed treatment for the control of seedling disease of water-sown rice. Hilgardia 41:689-698.

(Prepared by M. C. Rush)

Seedling Blight

Seedling blight occurs on dry-seeded rice (either drilled or broadcast) throughout the world. The disease is caused by seedborne and soilborne fungi and is greatly influenced by environmental conditions.

Symptoms

Seedling blight includes preemergence and postemergence death of seedlings and causes irregular or thin stands. Preemergence seedling blight is characterized by a discoloration or rot of the embryo area of the seed, resulting in no germination or destruction of the germinating embryo. Sometimes hyphal strands may be visible over and surrounding the rotting seed. If germination occurs, dark lesions may form on the radicle, coleoptile, or mesocotyl (Plate 27). These lesions may prevent emergence of the seedling from soil, or the seedling may emerge but be stunted or weakened. In postemergence seedling blight, dark brown or reddish brown lesions are usually present on the coleoptile or on primary or secondary leaves through about the three-leaf growth stage. Above the lesions, the leaves become blighted and die, turning straw-colored, light brown, or gray. Seedlings may suddenly blight and die without visible lesions.

Seedling blight caused by *Sclerotium rolfsii* Sacc. is typified by seedlings dying or turning white in patches or in short strips along a drill row. Fluffy white mycelium and small, round, tan sclerotia may be present on the soil surface at the base of affected seedlings.

Causal Organisms

Seedling blight, or damping-off, is a disease complex caused by several seedborne and soilborne pathogens. Among the seedborne fungi causing reduced germination, seed rotting, lesions on roots and coleoptiles, stunting, yellowing, and seedling blighting in water-seeded rice are *Alternaria padwickii* (Ganguly) Ellis, *Cochliobolus miyabeanus* (Ito & Kuribayashi in Ito) Drechs. ex Dastur (anamorph *Bipolaris oryzae* (Breda de Haan) Shoemaker), *Curvularia lunata* (Wakk.) Boedijn, *Pyricularia oryzae* Cavara, *Sarocladium oryzae* (Sawada) W. Gams & D. Hawksworth, *Magnaporthe salvinii* (Cattaneo) R. Krause & R. K. Webster (anamorph *Nakataea sigmoidea* (Cavara) K. Hara), *Gerlachia oryzae* (Hashioka & Yokogi) W. Gams, and *Fusarium moniliforme* J. Sheld. These fungi also cause foliage diseases and/or grain discoloration (pecky rice). The soilborne fungi *Pythium dissotocum* Drechs., *Pythium spinosum* Sawada, *Rhizoctonia solani* Kühn, and *Sclerotium rolfsii* Sacc. have been isolated from blighted seedlings in dry-seeded (drill-seeded or dry broadcast) rice. *S. rolfsii* can also cause a serious blighting of seedlings in upland nursery beds throughout Asia and in *dapog* beds in the Philippines.

Disease Cycle and Epidemiology

The seedling blight fungi survive between crops on seeds or in the soil. Damage caused by the pathogens is usually related to environmental conditions that inhibit the rapid germination and growth of rice. Low temperatures at the seedling stage, drought or lack of moisture from irrigation, and low soil fertility are all conditions that slow seedling development and favor seedling blight.

Seedling blight caused by *Pythium* spp. occurs when the soil is saturated but not flooded. Blighting caused by *S. rolfsii* is severe under warm, cloudy, humid conditions either in irrigated upland rice nurseries or in dry-seeded rice. Maximum disease development occurs around 30°C, but disease develops over the range 25–35°C. The fungus survives between crops as sclerotia. The sclerotia float and can be disseminated in irrigation water. Seedling blight caused by *R. solani* occurs under cold, wet conditions.

Control

Seedling blight is mainly controlled by the use of seed-protectant fungicides.

Both seed treatment and chemical soil treatment have controlled seedling blighting caused by *S. rolfsii* in *dapog* beds and in irrigated upland beds.

Avoiding environmental conditions unfavorable for seed germination and seedling growth minimizes problems with seedling blights. Flooding often stops further development of the seedling blight diseases, especially that caused by *S. rolfsii*.

Selected References

Anonymous. 1987. Rice production handbook. La. State Univ. Agric. Cent. Bull. 2321. pp. 36-46.

Groth, D. E., Rush, M. C., and Hollier, C. A. 1991. Rice diseases and disorders in Louisiana. La. Agric. Exp. Stn. Bull. 828.

Kuniyasu, K. 1983. The present situation of seed-borne diseases and seed disinfection in Japan. Jpn. Pestic. Inf. 43:13-22.

Ou, S. H. 1985. Rice Diseases. 2nd ed. Commonwealth Mycological Institute, Kew, England.

(Prepared by M. C. Rush)

Seed Box Diseases

The increased use of mechanical transplanting of rice during the 1970s and 1980s in Japan, Taiwan, and the Philippines has necessitated raising seedlings in nursery boxes or specialized beds (Plate 28). In the boxes, thick layers of seeds are kept under high temperature and moisture for several days until the seeds germinate. The boxes are then transferred to field nurseries, often covered with polyethylene sheets, where the seedlings are grown to transplant size. Many seedborne and soilborne pathogens are known to cause damping-off or seedling blight in the seed boxes.

Symptoms

Symptoms of damping-off and seedling blight in seed boxes range from poor germination and dark lesions on coleoptiles or mesocotyls to complete blighting of seedlings. Seedling blight may cause stunting and yellowing of seedlings and/or sudden blighting, with seedlings turning necrotic and straw-colored, brown, or gray (Plate 29). Roots are often discolored and rotted. The diseases tend to occur in circular areas within a box of seedlings. Cottony, white mycelium may be observed growing over the seeds, seedlings, and soil or planting medium. Bakanae disease, caused by *Fusarium moniliforme* J. Sheld., is expressed as abnormal elongation of the seedlings. Infected seedlings are several inches taller than healthy seedlings and are thin, with long, narrow, yellow-green leaves.

Causal Organisms

The causes of seedling blight in seed boxes include *Pyricularia grisea* (Cooke) Sacc., *Cochliobolus miyabeanus* (Ito & Kuribayashi in Ito) Drechs. ex Dastur, *Gibberella fujikuroi* (Sawada) Ito in Ito & K. Kimura (anamorph *Fusarium moniliforme*), *Phoma exigua* Desm., *Fusarium* spp., *Pythium* spp., *Rhizopus* spp., *Trichoderma viride* Pers.:Fr., *Rhizoctonia solani* Kühn, *Sclerotium rolfsii* Sacc., *Mucor fragilis* Bainier; the bacteria *Pseudomonas glumae* Kurita & Tabei, *Pseudomonas plantarii* Azegami et al, and *Pseudomonas alboprecipitans* Rosen; and the nematode *Aphelenchoides besseyi* Christie.

Disease Cycle and Epidemiology

These pathogens are mostly seedborne, although the *Pythium* spp. and *R. solani* can be soilborne. Seed boxes are heavily seeded on top of soil or other planting mixtures, and the fungal and bacterial pathogens spread rapidly under the warm, humid environmental conditions. Fungi such as *Rhizopus* spp. and *Trichoderma* spp., normally considered weak parasites, can cause considerable damage under the environmental conditions present in the seed boxes.

Control

Seed treatment with an appropriate fungicide has been effective in controlling these diseases. Also, treatment of the soil in seed boxes with chemical sterilants has significantly reduced seed box diseases.

Selected References

Kuniyasu, K. 1983. The present situation of seed-borne diseases and seed disinfection in Japan. Jpn. Pestic. Inf. 43:13-22.

Ohata, K., Umehara, Y., and Ibaragi, T. 1978. Diseases of Rice Seedlings Grown in Boxes and Methods for Preventing These Diseases. Nasangyoson-bunkakyokai, ed. Takeda Chemical, Tokyo.

Ou, S. H. 1985. Rice Diseases. 2nd ed. Commonwealth Mycological Institute, Kew, England.

(Prepared by M. C. Rush)

Foliar Diseases

Blast

Blast is one of the most important diseases of rice worldwide. Crop management greatly affects the disease. Losses and added costs for disease control are greatest where farmers practice intensive, high-input rice cultivation. In low-input, traditional agriculture, low soil levels of nitrogen limit disease development. Blast will likely become a common problem wherever technological and social change allow increased fertilizer use. Thus, blast is a constraint to the adoption of improved production practices.

Blast is most severe in irrigated rice grown in temperate regions or at high elevations in the tropics, and in rainfed, upland rice. Blast is least severe in tropical lowland rice grown with good water control, but it is a continuing problem where blast-susceptible cultivars are widely grown.

Symptoms

The pathogen produces lesions on all parts of the shoot but rarely attacks the leaf sheath. The characteristics of lesions on the leaf blades vary with environment and host resistance. When lesions first appear, they are often white to gray-green with darker green borders (Plate 30). Older lesions are generally whitish to gray with necrotic borders (Plate 31). Their shape varies, but lesions are characteristically diamond-shaped (Plates 32 and 33).

Lesion size varies with plant age and resistance. Necrotic spots are common on cultivars with major genes for resistance, but these spots may also be found intermixed with larger lesions on susceptible cultivars. Lesions on the leaf blades of plants at the reproductive stage are generally larger (2 cm long) than lesions on younger plants (less than 1 cm long).

Infection of other parts of the shoot can occur at the leaf collar, at the culm and culm nodes, and often at the panicle neck node (Plates 34 and 35). Collar infection can kill the entire leaf blade. Internodal infection of the culm can occur in a banded pattern, with 3 cm of blackened necrotic culm and 3 cm of healthy tissue in succession. In some upland rice environments, internodal infection occurs at the base of the plant and at heading causes a white-panicle symptom similar to that induced by stem-boring insects or water deficit. Nodal infection is more common than internodal infection, and the culm can break at the infected node.

Infection of the panicle neck node, called neck blast or rotten neck blast, is the most destructive symptom in most environments where blast is a problem (Plate 36). Neck node infection and infection of the panicle branches and spikelet pedicels may occur together, or they may occur separately under some conditions.

Although the biochemical basis of the pathogenicity of the fungus is not clear, toxins are likely components of pathogenicity. A number of toxins from the pathogen have been identified, including γ-picolinic acid, tyrosol, two 3,4-dihydroisocoumarins, several napthalenones, pyriculol, and tenuazonic acid.

Causal Organism

Pyricularia grisea (Cooke) Sacc. is the cause of rice blast. The name *P. oryzae* Cavara has been used widely in the phytopathological literature, but for the fungus attacking rice, *P. oryzae* is morphologically indistinguishable from *P. grisea*, which attacks other grasses. Both have the same teleomorph. Since these fungi are regarded as synonymous, the earlier name, *P. grisea*, is the correct name for the rice blast pathogen.

The teleomorph, *Magnaporthe grisea* (T. T. Hebert) Yaegashi & Udagawa, has not been found in nature. It is an ascomycete in the Physosporelleaceae that produces hyaline, fusiform, three-septate ascospores in unitunicate asci. The fungus is heterothallic with bipolar mating control, but there are additional genes controlling the sexual cycle. Nearly all rice field isolates are males and are unable to cross with one another, whereas many isolates from other grasses are hermaphroditic. Hermaphroditic laboratory strains able to infect rice have been developed through crossing with nonrice isolates.

After incubation of infected leaves in a moist chamber, conidiophores bearing conidia develop as a fleecy gray growth over the lesion surface. The conidia have mucilage encased in a periplasmic space at the spore tip (Fig. 5A–C). When the mucilage is released, it adheres strongly to hydrophobic surfaces and may be a unique mechanism to attach conidia to leaves.

Disease Cycle and Epidemiology

Blast is favored by culture of rice in aerobic soil or in wetlands where intermittent drought stress occurs, by N fertilization, by long duration of leaf wetness due to drizzle or dew deposition, by little or no wind at night, and by night temperatures between 17 and 23°C.

The pathogen overseasons in infested crop residue or in seed in temperate regions where only one rice crop is sown per year. Seedborne inoculum is an important source of primary infection where seed is sown densely in seedling boxes for mechanical transplanting. Seedborne inoculum may also be important in arid climates, for example, in Egypt, but is unimportant in the humid tropics, where other crops of rice and crop residue serve as inoculum sources. Weed hosts may also be a source of primary inoculum, but host-range studies conducted in greenhouses may not accurately reflect susceptibility in the field. Little is known of the importance of inoculum from weed hosts under natural conditions.

Conidia are produced and released during periods of high relative humidity, and daily production generally peaks between midnight and 0600 hr. In the tropics, conidia have been trapped during daytime rain showers. No sporulation occurs below 89% relative humidity, and sporulation increases with increasing relative humidity above 93%. The optimum temperature for sporulation is 25–28°C. The highest rate of spore production is attained 3–8 days after lesion appearance, and lesions maintain the ability to produce spores longer at 16–24°C than at 28°C.

Leaf wetness (i.e., free water) is essential for infection. Optimum temperatures are 25–28°C for germination and 16–25°C for appressorium formation. At 15.6°C, leaf wetness for 16–20 hr is required for infection; at 24.4°C, 8–11 hr is required. Under experimental conditions, the pathogen can infect within 6 hr at 24°C. The latent period required for lesion formation varies from 13 to 18 days at 9–11°C to an optimum of 4–5 days at 26–28°C.

Resistance and Pathogen Variability

Complete resistance to blast, where the fungus is unable to cause sporulating lesions on the plant, is well known, but it also has been associated with spectacular breakdowns in cultivar resistance. This type of resistance is controlled by one or two genes and is generally effective for 2 or 3 years in environments conducive to disease. At least 13 genes for complete resistance have been identified. After the complete resistance of a cultivar has been overcome, the cultivar may exhibit either high susceptibility or some level of residual resistance.

Quantitative resistance of cultivars that are qualitatively susceptible has been variously referred to as slow blasting,

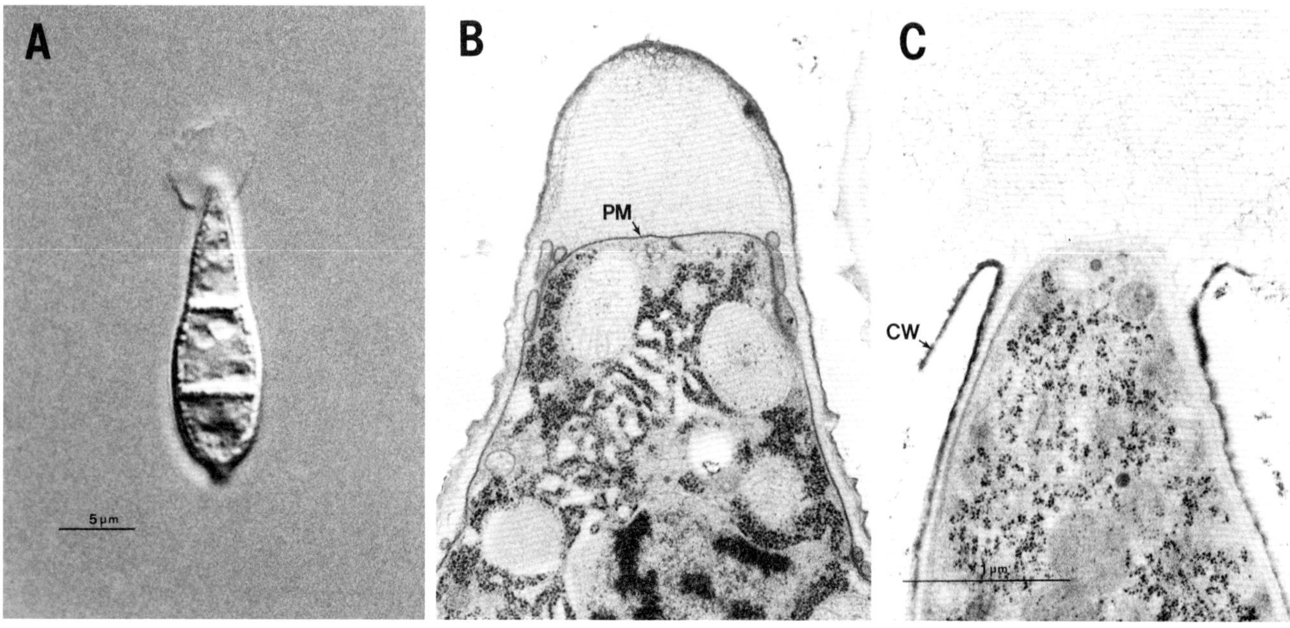

Fig. 5. Light (**A**) and electron (**B** and **C**) micrographs of *Pyricularia grisea* conidia before (**B**) and after (**A** and **C**) hydration. **A,** Conidium imaged with differential interference contrast optics. **B,** A dry conidium apex contains a deposit outside the plasma membrane (PM). **C,** When hydrated, the cell wall (CW) is broken at the apex and an extracellular fibrous matrix is released. (Reprinted, by permission, from Hamer et al, 1988; © AAAS)

or as dilatory, field, or partial resistance. This type of resistance has been shown to be durable in farm fields, especially in many tropical lowland environments of Asia. Cultivars with partial resistance have fewer and smaller lesions than fully susceptible cultivars, and the lesions sporulate less. A latent period, however, apparently is not a key component of partial resistance to blast.

Some forms of partial blast resistance are inherited monogenically and are race-specific, such as the resistance conferred by the *Pi-f* gene. However, the partial resistance in cultivars that have proven durably resistant is due to an undetermined number of genes.

Blast is a disease of young host tissues. For example, fully extended leaves are more resistant than partially extended leaves, leaves on older plants are more resistant, and panicle resistance increases with age.

Temperature affects resistance. Although high temperature (above 28° C) favors the growth of *P. grisea*, it also stimulates host resistance.

Most rice cultivars that are susceptible at the seedling stage are also susceptible to neck infection. Some cultivars exhibit adult-plant resistance and may be relatively resistant to leaf and neck infection at the reproductive stage. Consequently, seedling tests alone are inadequate to fully describe the resistance of some cultivars, and evaluation at later growth stages is required.

Traditionally, nurseries sown in aerobic soil have been used to assess resistance to leaf infection at the seedling stage. Complete resistance is easily identified by this nursery technique. However, the technique is not adequate for assessing the relative progress of the disease. For this purpose, a modification of the traditional nursery method has been developed using three-row miniplots with a barrier cultivar on three sides and an inoculum donor row on one side. This method is convenient for screening a large number of entries and can be used to screen pedigree lines as well.

The degree of variation in *P. grisea* has been debated in the literature. Some researchers concluded that sister isolates from single lesions or single monospore cultures consist of many races, as identified on differential sets of cultivars. They suggested that if enough differential cultivars were used, possibly every conidium from a culture could be shown to represent a different race.

In contrast, other studies indicated that the fungus is more stable. Much of the variation encountered in the earlier work could have been due to undetected variation in plant physiological condition and environment. Current evidence indicates that although there are many races of *P. grisea* and may be pathogenically unstable variants, the organism does not appear to generate many new races with each asexual reproduction cycle.

Control

In environments that are not highly conducive to blast, the disease can be controlled most easily by planting resistant cultivars. In fact, satisfactory control in many irrigated areas in the tropics has been achieved with resistance alone. However, in environments with high epidemic potential, and where few durably resistant cultivars are available, it is necessary to combine other means of blast control with host resistance. These include manipulating time of planting, fertilizer, and water management as well as the use of fungicides.

In tropical upland rice, crops sown early after the onset of the rainy season are more likely than late-sown crops to escape blast infection. In the upland rice areas of Brazil, farmers are advised to sow early to escape inoculum originating from neighboring farms. In Bangladesh, blast is most severe during seasonal periods of low night temperatures and long dew duration. Early sowing may limit exposure of the crop to these conditions.

Nitrogen fertilizer application strongly increases the susceptibility of rice to blast; phosphorus fertilization may also increase susceptibility in certain soils. A limit of 15 kg N per hectare is recommended for upland rice in Brazil to reduce vulnerability to blast. N form also affects blast severity. In nutrient solution culture, plants given NO_3^- are more susceptible than plants given NH_4^+. This effect of N form may partially explain why plants grown in aerobic soil, where NO_3^- is the main source of inorganic N, are more susceptible than plants grown in flooded soil, where NH_4^+ is the main inorganic N source. Research in Japan has shown that split application of N reduces blast compared with a single application, but the ideal timing, number of splits, and quantity of N vary with soil and climatic conditions. Under highly blast-conducive conditions in the Philippines, timing alone was not an effective means of control. To develop fertilizer management practices that will contribute to blast control without loss of yield or profit, a more fundamental knowledge of the interaction between disease and host mineral nutrition is necessary.

Drought stress also increases blast susceptibility; water management practices that lessen the likelihood of stress aid in the control of blast in rainfed areas.

Fungicides are often used for blast control in industrialized countries. The development of highly effective systemic fungicides, such as pyroquilon and tricyclazole, has opened new possibilities for controlling the disease in developing countries. These compounds have such high activity that seed treatments may result in long-term control of leaf blast. If seed treatments are effective, fungicides may be economical in many developing countries. However, chemicals should always be viewed as one component of an integrated control approach adapted to local needs. Blast forecasting may open the possibility of more rational use of fungicides in areas where chemical control is currently practiced. Also, blast simulation models might prove useful in predicting the potential for the disease in regions where rice production practices are changing.

Selected References

Bonman, J. M., and Mackill, D. J. 1988. Durable resistance to rice blast disease. Oryza 25:103-110.

Bonman, J. M., Vergel de Dios, T. I., Bandong, J. M., and Lee, E. J. 1987. Pathogenic variability of monoconidial isolates of *Pyricularia oryzae* in Korea and in the Philippines. Plant Dis. 71:127-130.

Froyd, J. D., Paget, C. J., Guse, L. R., Dreikorn, B. A., and Pafford, J. L. 1976. Tricyclazole: A new systemic fungicide for control of *Piricularia oryzae* on rice. Phytopathology 66:1135-1139.

Hamer, J. E., Howard, R. J., Chumley, F. G., and Valent, B. 1988. A mechanism for surface attachment of spores of a plant pathogenic fungus. Science 239:288-290.

Hebert, T. T. 1971. The perfect stage of *Pyricularia grisea*. Phytopathology 61:83-87.

Hwang, B. K., Koh, Y. J., and Chung, H. S. 1987. Effects of adult-plant resistance on blast severity and yield of rice. Plant Dis. 71:1035-1038.

Kahn, R. P., and Libby, J. L. 1958. The effect of environmental factors and plant age on the infection of rice by the blast fungus, *Piricularia oryzae*. Phytopathology 48:25-30.

Kingsolver, C. H., Barksdale, T. H., and Marchetti, M. A. 1984. Rice blast epidemiology. Bull. Pa. Agric. Exp. Stn. 853.

Kürschner, E. M., Bonman, J. M., Garrity, D. P., Neue, H. U., Tamisin, M. M., Estrada, B. A., and Quijano, C. C. 1991. Effects of some cultural practices on blast disease and rice yield at an acid upland site. Proc. Int. Conf. Plant Prot. Trop. 3rd.

Latterell, F. M. 1975. Phenotypic stability of pathogenic races of *Pyricularia oryzae*, and its implication for breeding of blast resistant varieties. Pages 199-234 in: Horizontal Resistance to the Blast Disease of Rice. Ser. CE-9. Centro Internacional de Agricultura Tropical, Cali, Colombia.

Latterell, F. M., and Rossi, A. E. 1986. Longevity and pathogenic stability of *Pyricularia oryzae*. Phytopathology 76:231-235.

Levy, M., Ramao, H., Marchetti, M. A., and Hamer, J. E. 1991. DNA finger printing with a dispersed repeated sequence resolves pathotype diversity in the rice blast fungus. Plant Cell 3:95-102.

Mackill, A. O., and Bonman, J. M. 1986. New hosts of *Pyricularia oryzae*. Plant Dis. 70:125-127.

Marchetti, M. A. 1983. Dilatory resistance to rice blast in USA rice.

Phytopathology 73:645-649.
Ou, S. H. 1985. Rice Diseases. 2nd ed. Commonwealth Mycological Institute, Kew, England.
Ou, S. H., Nuque, F. L., Ebron, T. T., and Awoderu, V. A. 1971. A type of stable resistance to blast disease of rice. Phytopathology 61:703-706.
Prabhu, A. S. and Morais, O. P. 1986. Blast disease management in upland rice in Brazil. Pages 383-392 in: Progress in Upland Rice Research. International Rice Research Institute, Manila, Philippines.
Rossman, A. Y., Howard, R., and Valent, B. 1990. *Pyricularia grisea*, the correct name for the rice blast disease fungus. Mycologia 82:509-512.
Suzuki, H. 1975. Meteorological factors in the epidemiology of rice blast. Annu. Rev. Phytopathol. 13:239-258.
Teng, P. S., Klein-Gebbinck, H., and Pinnschmidt, H. 1991. An analysis of the blast pathosystem to guide modeling and forecasting. Pages 1-30 in: Rice Blast Modeling and Forecasting. International Rice Research Institute, Manila, Philippines.
Yaegashi, H., and Udagawa, S. 1978. The taxonomical identity of the perfect state of *Pyricularia grisea* and its allies. Can. J. Bot. 56:180-183.
Yeh, W. H., and Bonman, J. M. 1986. Assessment of partial resistance to *Pyricularia oryzae* in six rice cultivars. Plant Pathol. 35:319-323.

(Prepared by J. M. Bonman)

Brown Spot

Brown spot occurs on upland and wetland rice in all major rice-growing regions of the world. An epidemic of the disease in Bengal in 1942 resulted in yield losses of 40–90% and was largely responsible for the Bengal famine of 1943.

Symptoms

Brown spot may be manifested as a seedling blight or as a foliar and glume disease of mature plants (Plate 37A and B). On seedlings, the fungus produces small, circular, brown lesions, which may girdle the coleoptile and cause distortion of the primary and secondary leaves. In some cases, the fungus may also infect and cause a black discoloration of the roots. Infected seedlings are stunted or killed. On the leaves of older plants, the fungus produces circular to oval lesions that have a light brown to gray center surrounded by a reddish brown margin. On moderately susceptible cultivars, the lesions are 1–4 mm long, whereas on susceptible and highly susceptible cultivars, the lesions are 5–14 mm long. On resistant cultivars, the fungus produces tiny, dark specks. When infection is severe, the lesions may coalesce, killing large areas of affected leaves. The fungus also may infect the glumes, causing dark brown to black oval spots, and may also infect the grain, causing a black discoloration.

Brown spot infection reduces the number of grains per panicle and the kernel weight.

Causal Organism

Brown spot is caused by *Cochliobolus miyabeanus* (Ito & Kuribayashi in Ito) Drechs. ex Dastur, anamorph *Bipolaris oryzae* (Breda de Haan) Shoemaker (syns. *Dreschlera oryzae* (Breda de Haan) Subramanian & P. C. Jain and *Helminthosporium oryzae* Breda de Haan).

Ascomata are dark brown and globose to elliptical, 360–780 μm in diameter, with a conical to cylindrical beak (98–200 × 55–110 μm). Asci (140–235 × 21–26 μm) are clavate, cylindrical-clavate, or broadly fusoid; straight to somewhat curved; and bitunicate. Ascospores (235–648 × 4–9 μm) are hyaline to light olivaceous brown, filiform or flagelliform, wider in the middle and attenuated towards the ends, eight- to 12-septate, surrounded by a mucilaginous sheath, and coiled in a helix within the ascus.

Conidiophores are single or in groups, straight to flexuous, sometimes geniculate, pale brown to olivaceous brown, lighter toward the apex, septate, up to 600 μm long, and 4–8 μm wide. Conidia are fusoid, obclavate to sometimes almost cylindrical, generally curved, light brown to golden brown, six- to 14-distoseptate, and 63–153 × 14–22 μm, often with a slightly protruding hilum that may be light or dark. The conidia usually germinate from one or both polar cells.

The fungus produces a non-host-specific toxin called ophiobolin.

Disease Cycle and Epidemiology

In temperate regions, the primary source of inoculum is infected and infested seed, but the fungus also overwinters in infected crop debris. In the tropics, seedborne inoculum may be less important, because inoculum is present year-round in overlapping rice crops. Species of some 23 grass genera and 20 species of *Oryza* have been reported to be susceptible to infection by *B. oryzae* in artificial inoculations, but reports of natural infection of nonrice hosts are few, except for *Zizania aquatica* L. (commercial wild rice). The role of ascospores in the disease cycle is not known, and there is no solid evidence of pathogenic specialization by the fungus.

Infection can occur between 16 and 36° C; optimum temperatures are reported to lie between 20 and 30° C. Although infection is favored by free water on plant surfaces, it has been reported without free water at relative humidities above 89%. Water stress increases host susceptibility.

Brown spot is most prevalent on rice that is grown in nutrient-deficient soils, specifically soils low in potassium, manganese, magnesium, silica, iron, and calcium. Both high and low levels of nitrogen have been reported to increase susceptibility of rice plants to brown spot. Some researchers think that yield losses associated with the disease result principally from the nutrient deficiencies that predispose the plants to the disease rather than to the disease itself. Rice grown in unflooded soil is also more susceptible to brown spot.

Climatic factors that were correlated with the brown spot epidemic in Bengal in 1942 were above-normal rainfall, high humidity, prolonged periods of cloudy weather, and warmer-than-average temperatures during the flowering and grain-filling stages.

Control

The most important factors in control of brown spot are proper nutrients in the soil and prevention of water stress. Brown spot is more an indicator of nutritional or physiological disorders than a pathological disease. In areas where the disease is severe and associated soil abnormalities are not easily corrected, resistant cultivars should be introduced or developed. Chemical seed treatment has proven effective in reducing seedling disease caused by *B. oryzae*. Foliar applications of fungicides to reduce secondary infection are of doubtful practical value.

Selected References

Aluko, M. O. 1970. The measurement of brown leaf spot on rice. PANS 16:76-81.
Klomp, A. O. 1977. Early senescence of rice and *Drechslera oryzae* in Wageningen Polder, Surinam. Versl. Landbouwkd. Onderz. 859.
Marchetti, M. A., and Petersen, H. D. 1984. The role of *Bipolaris oryzae* in floral abortion and kernel discoloration in rice. Plant Dis. 68:288-291.
Ou, S. H. 1985. Rice Diseases. 2nd ed. Commonwealth Mycological Institute, Kew, England.
Padmanabhan, S. Y. 1973. The great Bengal famine. Annu. Rev. Phytopathol. 11:11-26.
Sivanesan, A. 1987. Graminicolous species of *Bipolaris, Curvularia, Drechslera, Exserohilum*, and their teleomorphs. Commonw. Mycol. Inst. Mycol. Pap. 158.

(Prepared by F. N. Lee)

Leaf Scald

Leaf scald, also known as leaf tip blight, leaf tip dying, brown leaf blight, and leaf sheath browning, occurs in the southern United States, Latin America, and West Africa and throughout Asia. The disease has the potential to significantly lower yields.

Symptoms

A variety of symptoms are associated with leaf scald. The classic leaf symptom is zonate lesions of alternating light tan and dark brown that begin at the leaf edges or tips (Plate 38). Lesions may coalesce and destroy large areas of the leaf (Plate 39). As the lesions mature, the zonate pattern fades, and affected areas dry out, giving the leaf a scalded appearance. In some countries, lesions rarely develop the zonate pattern and only the scalding symptom is evident. Other symptoms caused by the disease include small, reddish brown spots on the leaves and patches of brown lesions on the sheaths. In Costa Rica, the disease has been reported to cause a panicle blight, resulting in deformed or sterile flowers and light brown discoloration of the glumes. Seed decay has also been reported.

Causal Organism

The anamorph, *Gerlachia oryzae* (Hashioka & Yokogi) W. Gams (syn. *Rhynchosporium oryzae* Hashioka & Yokogi), produces sporodochia with masses of pink to salmon conidia. The conidia are falcate, 6–15 × 3–4.5 μm, nonseptate or with up to three septa (usually one), and produced in slime from annellidic conidiogenous cells. In culture, colonies are initially white, becoming cream, pink, or salmon with age. Optimum growth occurs at 24–28°C. The teleomorph, *Monographella albescens* (Thuem.) Parkinson et al (syn. *Metasphaeria albescens* Thuem.), produces brown, globose perithecia that are embedded in the leaf tissue, except for the ostiole. Asci are cylindrical-clavate and unitunicate and have a distinct apical ring that blues in iodine. Ascospores are fusoid, straight or somewhat curved, three- to five-septate, 14–23 × 3.5–4.5 (–7.5) μm, and obliquely distichous to tristichous.

Disease Cycle and Epidemiology

The fungus survives on rice seed and in dry crop debris. *Echinochloa crus-galli* (L.) P. Beauv. is an alternate host for the fungus and may serve as a source of inoculum. Disease development usually occurs late in the season on mature leaves and is favored by wet weather and high nitrogen fertilization. The fungus enters the host through the stomates.

Selected References

Hashioka, Y., and Ikegami, H. 1955. The leaf scald of rice. Pages 46-51 in: Jubilee Publication Commemorating the 60th Birthday of Professors Y. Tochiorai and T. Fukushi. Sapporo, Japan.

Hashioka, Y., and Makine, M. 1956. Relation of nitrogen nutrition of rice plants to the susceptibility to four foliage diseases. Res. Bull. Fac. Agric. Gifu Univ. 6:58-66.

Parkinson, V. O., Sivanesen, A., and Booth, C. 1981. The perfect state of the rice leaf scald fungus and the taxonomy of both the imperfect and perfect states. Trans. Br. Mycol. Soc. 76:59-69.

Shanmughon, S. N., Gopalar, N., George, K. M., and Gopalakuishnan, R. 1973. Leaf scald of rice. Curr. Sci. 42:582-583.

(Prepared by D. Groth)

Narrow Brown Leaf Spot

Narrow brown leaf spot, also known as Cercospora leaf spot, was first described in Japan in 1906. The disease has been reported from major rice-growing regions in the tropics and subtropics of Asia, Africa, Australia, and North, Central, and South America. Yield losses of up to 40% have been reported.

Symptoms

Symptoms usually occur during the late growth stages and are characterized by short, elliptical to linear brown lesions on the leaf blades, 210 × 11.5 mm (Plate 40). The lesions tend to be narrow, short, and dark brown on resistant cultivars and wider and lighter brown with gray necrotic centers on susceptible cultivars. Leaf necrosis may occur on susceptible cultivars. Symptoms similar to those on the leaves also occur on the leaf sheaths, pedicels, and glumes. A net blotch-like pattern often forms on leaf sheaths, where the cell walls turn dark brown and the intercellular areas are tan to yellow. The disease can cause premature ripening of kernels and lodging of plants.

Causal Organism

Narrow brown leaf spot is caused by *Cercospora oryzae* Miyake. The fungus produces pale brown to dark brown, geniculate, multiseptate (three or more) conidiophores, singly or in fascicles (generally 17), which emerge from the host stomates and measure 80–140 × 4–9 μm. Conidia are hyaline, cylindrical to narrowly clavate, three- to 10-septate, and 15–60 × 3.6–4.0 μm. *Sphaerulina oryzina* K. Hara was described in 1918 as the teleomorph of *C. oryzae*.

Optimum growth of the fungus occurs at 25–28°C. Several races of *C. oryzae* have been identified based on the reactions of a differential set of eight cultivars.

Disease Cycle and Epidemiology

C. oryzae penetrates the host through the stomata and grows intercellularly in the host tissue. Symptoms may take up to 30 days to develop after infection. The primary factors affecting disease development are susceptibility of the cultivar to the race or races of the fungus present and the growth stage of the plant. Although rice plants are susceptible to the fungus at all stages of growth, they are most susceptible from panicle emergence to maturity. In addition to *O. sativa*, the fungus is also reported to infect *Panicum repens* L.

Control

The disease is controlled through the use of resistant cultivars.

Selected References

Estrada, B. A., and Ou, S. H. 1978. Methods of screening rices for varietal resistance to Cercospora leaf spot. IRRI Res. Pap. Ser. 19.

Ou, S. H. 1985. Rice Diseases. 2nd ed. Commonwealth Mycological Institute, Kew, England.

Sah, D. N., and Rush, M. C. 1988. Physiological races of *Cercospora oryzae* in the southern United States. Plant Dis. 72:262-264.

Sridhar, R. 1970. A new collateral host of *Cercospora oryzae*. Plant Dis. Rep. 54:272.

(Prepared by C. Hollier)

Stackburn

Stackburn was first described in the southern United States and subsequently has been found in rice-growing regions worldwide.

Symptoms

The causal fungus produces oval to circular spots on the leaves, 3–10 mm in diameter, which are initially tan, becoming gray to white, and are surrounded by a narrow, dark brown border (Plate 41). Small black dots (sclerotia) are produced in the centers of older lesions. The fungus produces lesions on the glumes that are similar, except with wider borders. Infection of the kernels produces discolored, shriveled, or brittle grains. Dark brown lesions may also occur on the coleoptiles of germinating seeds or young seedlings. Severely diseased seedlings are killed.

Causal Organism

Stackburn is caused by *Alternaria padwickii* (Ganguly) Ellis (syn. *Trichoconis padwickii* Ganguly). The conidia are yellow to pale brown, fusiform to obclavate, rostrate, and smooth or minutely echinulate, with three to five transverse septa. The beak is hyaline and may be septate. Conidia are 95–170 μm long (including the beak) and 11–20 μm wide. Conidiophores are 100–180 $\times$ 3–4 μm, often swollen at the apex, and produce conidia singly. Sclerotia are black, spherical, and reticulate, measuring 50–220 μm in diameter.

Disease Cycle and Epidemiology

The fungus survives in soil, on seed, and in infested plant debris as sclerotia and mycelia. Infection occurs via airborne conidia. Infection normally occurs through wounds, but the pathogen may also directly penetrate the glumes and infect immature kernels. Seed infection often leads to infection of the seedling, causing a seedling blight.

Control

No specific control recommendations are available, but seed-protectant fungicides and hot-water treatment reduce the incidence and severity of seedling blight. Some differences in resistance among cultivars have been noted.

Selected References

Chuaiprasit, C. 1976. Morphology, physiology, and pathogenicity of *Trichoconis padwickii* Ganguly, the cause of stackburn disease of rice. Ph.D. thesis. Oregon State University, Corvallis.

Mathur, S. B., Mallya, J. I., and Neerguard, P. 1972. Seed-borne infection of *Trichoconis padwickii* in rice, distribution and damage to seeds and seedlings. Proc. Int. Seed Test. Assoc. 37:803-810.

Tisdale, W. H. 1922. Seedling blight and stackburn of rice and the hot water seed treatment. Bull. U.S. Dep. Agric. 1116.

(Prepared by D. Groth)

Leaf Smut

Leaf smut is found in most rice-growing regions of the world and is considered to be a minor disease.

Symptoms

The fungus produces small, slightly raised, black spots or streaks (sori) under the epidermis of leaves and sometimes on the leaf sheaths. The sori are linear to rectangular, 0.5–5.0 $\times$ 0.25–1.5 mm, and run parallel to the veins (Plate 42). Under heavy infestation, the leaves turn yellow, the tips dry out and die, and the leaves split lengthwise.

Causal Organism

Leaf smut is caused by *Entyloma oryzae* Syd. & P. Syd. (syns. *E. lineatum* (Cooke) J. J. Davis and *E. dactylidis* (Pass.) Cif.). The fungus produces sori consisting of agglutinated teliospores that are angular-globose, pale brown alone and black in mass, and 7–11 μm in diameter. The spores germinate to form a promycelium, which at its tip produces three to seven narrowly elliptical, hyaline sporidia, 10–15 $\times$ 2–2.5 μm. Secondary sporidia develop from the apex of primary sporidia, forming Y-shaped structures.

Disease Cycle and Epidemiology

The fungus survives between rice crops as teliospores on diseased leaf debris in the soil. Sporidia from germinated teliospores are disseminated by wind to the rice leaves. Infection normally occurs late in the season. High nitrogen levels in the plant, especially during the late growth stages, favor disease development. The fungus also infects wild rice, *Zizania aquatica* L.

Control

Resistant cultivars provide the best method of control if disease control is warranted. Foliar fungicides have been shown to be effective in controlling leaf smut but are not economical.

Selected References

Singh, R. A., and Pavgi, M. S. 1960. Varietal reaction to leaf smut of rice. Indian Phytopathol. 19:378-382.

Taubenhaus, J. J. 1964. Research on rice disease in Arkansas. Rice J. 65:28.

Templeton, G. E. 1967. Rice research in Arkansas. Rice disease research. 1966. Rice J. 70:28-30.

Tullis, E. C. 1934. Leaf smut of rice in the United States. Phytopathology 24:1386.

(Prepared by D. Groth)

Eyespot

Eyespot of rice was first reported in 1980 in Colombia, Panama, and Peru, although the disease was known to occur on several other grasses and on banana. Subsequently, the disease was found on rice in Bolivia, Brazil, Venezuela, Guatemala, Mexico, Honduras, and Jamaica. The disease has not caused any serious yield losses to date.

Symptoms

Lesions occur on the leaves and can be found at all stages of growth. Initially, the lesions appear as small, oval, water-soaked, gray-green spots, 1–4 $\times$ 0.5–1 mm. A yellow halo frequently surrounds young lesions. As the disease progresses, the lesions enlarge and become white to straw-colored and delimited by a well-defined, brown margin (Plate 43). Under favorable conditions, successive development and coalescence of lesions produce a characteristic zonate pattern. The leaves of highly susceptible cultivars turn yellow and wither immediately after heading (Plate 44). Symptoms of eyespot are sometimes confused with those of brown spot, leaf blast, and leaf scald.

Causal Organism

Drechslera gigantea (Heald & F. A. Wolf) Ito produces dark gray or olivaceous colonies. Conidiophores are dark brown and one- to seven-septate, with a slightly swollen tip, measure 128–345 $\times$ 9.9–17 μm, and produce macroconidia singly at the tip. Macroconidia are cylindrical, with hemispherical or hemiellipsoidal ends, hyaline to pale yellow or light brown, usually three- to six-septate (sometimes eight), and 166–484 $\times$ 10–30 μm (average 330 $\times$ 19 μm). Macroconidia of the strain from wild rice (*Zizania aquatica* L.) differ by having five to 16 septa. Microconidia are hyaline to olive, one- to two-septate, ellipsoid, 9.5–24 $\times$ 3.5–7 μm (average 14 $\times$ 5 μm), and formed in chains.

Disease Cycle and Epidemiology

D. gigantea is reported to survive between rice crops on several grass hosts as dormant mycelium. The pathogen may also survive on volunteer rice. Barnyard grass, goosegrass, *Bromus* spp., reed canarygrass, quackgrass, velvetbean, and *Z. aquatica* are all reported as hosts. Cool night temperatures, high humidity, and extended periods of leaf wetness favor disease development.

Control

The control of eyespot on rice has not been seriously explored, because the pathogen has limited virulence on most commercial cultivars. Only certain cultivars derived from Mong Chin Vang A, a Vietnamese cultivar, have been found to be highly susceptible to the disease.

Selected References

Ahn, S.-W. 1980. Eyespot of rice in Colombia, Panama, and Peru. Plant Dis. 64:878-880.

Kardin, M. K., Bowden, R. L., Percich, J. A., and Nickelson, L. J. 1982. Zonate eyespot on wild rice caused by *Drechslera gigantea*. Plant Dis. 66:737-739.

Shoemaker, R. A. 1962. *Drechslera* Ito. Can. J. Bot. 40:809-836.

(Prepared by S.-W. Ahn)

Downy Mildew

Downy mildew of rice has been reported from Italy, India, China, Japan, Korea, Taiwan, and the United States. The disease is considered to be of minor importance.

Symptoms

The leaves of infected plants may have yellow to white spots or patches, and the flag leaf and upper leaves may be curled, bent, or twisted. The symptoms are most evident on the panicle, which fails to emerge completely, is contorted, remains green, and fails to produce any grain (Plate 45). The floral parts are generally abortive, and the lemma and/or palea exhibit phyllody and are converted into leafy structures.

Causal Organism

Downy mildew is caused by *Sclerophthora macrospora* (Sacc.) Thirumalachar, C. G. Shaw & Narasimhan (syn. *Sclerospora macrospora* Sacc.). The fungus has oospores that are typical of the genus *Sclerospora* and sporangia and sporangiophores that are typical of the genus *Phytophthora*. The sporangiophores are undifferentiated, simple, or sympodial, and emerge in groups through the host stomates. Sporangia are pyriform to ellipsoid, papillate, and 60–100 $\times$ 43–64 μm. Zoospores are biflagellate and ovate to kidney-shaped. Oogonia are globose, 46–80 $\times$ 42–69 μm, with a thick wall that is confluent with the wall of the oospore. Antheridia are obovate to ellipsoid, paragynous, and 20–48 $\times$ 10-29 μm. Oospores are near globose, 45–70 μm in diameter, with a thick wall (4–10 μm), and germinate to form sporangia.

Disease Cycle and Epidemiology

The fungus has a wide host range, infecting many gramineous hosts including oats, barley, wheat, maize, millet, and rye. The fungus survives as oospores in infected host debris in the soil. Oospores or sporangia from alternate hosts function as primary inoculum for infection of rice seedlings. Zoospores released from sporangia invade the plumule or infect the first leaves of seedlings, and the fungus grows systemically in the rice host. Sporulation and infection are favored by cool temperatures (18–23°C) and high humidity, and oospore germination is favored by free water. Disease incidence has been reported to be lowest when seedlings are grown under flood.

Control

Control measures are not generally warranted. Treating seed and covering nursery beds with a layer of fungicide-treated rice hulls have been reported to protect emerging rice seedlings from infection.

Selected References

Gangopadhyay, S. 1983. Current Concepts on Fungal Diseases of Rice. Today and Tomorrow's Printers & Publishers, New Delhi, India.

Ou, S. H. 1985. Rice Diseases. 2nd ed. Commomwealth Mycological Institute, Kew, England.

Thirumalacher, M. J., Shaw, C. G., and Narasimhan, M. J. 1953. The sporangial phase of downy mildew on *Eleusine coracana* with a discussion of the identity of *Sclerospora macrospora* Sacc. Bull. Torrey Bot. Club 80:299-307.

(Prepared by F. N. Lee)

White Leaf Streak

White leaf streak occurs in Papua New Guinea, the Solomon Islands, North Borneo, Sierra Leone, and Nigeria.

Symptoms

The disease is characterized by small, interveinal, oblong to linear lesions, 1–3 $\times$ 0.5 mm, visible on both leaf surfaces. The lesions are white to light gray and surrounded by a distinct, very narrow, brown margin. They may form long, uninterrupted white streaks on heavily infected plants. Young lesions may be white on the leaf surface and brown on the undersurface.

Causal Organism

White leaf streak is caused by *Mycovellosiella oryzae* (Deighton & Shaw) Deighton (syn. *Ramularia oryzae* Deighton & Shaw). Condiophores are short, separate, hyaline, erect, and 4–20 $\times$ 2 μm, with conspicuous conidial scars. The conidiophores are formed from external mycelium on the lesion surface, which emerges from the leaf stomates. Conidia are hyaline, straight, cylindrical, with somewhat tapered ends, aseptate or with up to three septa, have a conspicuous hilum, are sometimes catenulate, and measure 15–40 $\times$ 0.75–1.25 μm.

Disease Cycle and Epidemiology

No information on the epidemiology or control of this disease is available.

Selected References

Deighton, F. C. 1979. Studies on *Cercospora* and allied genera. VII. New species and redispositions. Commonw. Mycol. Inst. Mycol. Pap. 144.

Deighton, F. C., and Shaw, D. 1960. White leaf streak caused by *Ramularia oryzae* sp. nov. Trans. Br. Mycol. Soc. 48:516-518.

(Prepared by P. S. Gunnell)

Collar Rot

Collar rot is the name applied to a minor disease of rice in Thailand.

Symptoms

The disease occurs at the late tillering stage and is characterized initially by small, dark brown lesions at the base of the auricle, usually on the uppermost, fully expanded leaves. The lesions enlarge and extend to the leaf sheath and leaf blade. Eventually, the collar becomes rotted and the leaf blade drops off.

Causal Organism

Originally, the pathogen was identified as *Ascochyta oryzae* Cattaneo. Subsequently, it was identified as *Phomopsis oryzae-sativae* Punithalingam. Pycnidia of the fungus are erumpent, black, subglobose, ostiolate, unilocular or bilocular, and up to 550 μm in diameter. Conidiophores line the pycnidial cavity and are simple, hyaline, sometimes septate, phialidic, and up to 45 μm in length. α-Conidia are hyaline, one-celled, fusiform to ellipsoid, with two to three (sometimes four) guttae, and measure 11–14.5 × 2.5–3 (up to 3.5) μm. β-Conidia are absent.

Disease Cycle and Epidemiology

No information on the epidemiology or control of this disease is available.

Selected References

Ou, S. H. 1985. Rice Diseases. 2nd ed. Commonwealth Mycological Institute, Kew, England.

Punithalingam, E. 1980. *Phomopsis oryzae-sativae*, No. 665. Descriptions of Pathogenic Fungi and Bacteria. Commonwealth Mycological Institute, Kew, England.

(Prepared by P. S. Gunnell)

Rusts

Early reports of rusts on rice include *Puccinia graminis* f. sp. *oryzae* in Italy, Spain, and Romania and *Uromyces coronatus* Yosh in China. These reports have not been followed with additional information on their occurrence or of crop loss due to these pathogens.

P. graminis f. sp. *oryzae* was originally reported by Briosi in Italy. The validity of the report and identity of the pathogen were reviewed by Diehl. He concluded that, although there was considerable confusion in the early reports, the name *P. graminis* f. sp. *oryzae* was valid and that the fungus was morphologically identical to *P. graminis* on other cereals. An extensive search of the literature from recent years has not revealed additional reports of this disease on rice.

Uromyces coronatus Yosh was reported on *Oryza sativa* L. and *Zizania aquatica* L. (wild rice) from China in 1935. *U. coronatus* Miyabe & Nishida ex Dietel has since been reported in Florida and Asia on *Zizaniopsis* sp. No details of the nature and extent of the disease on rice were given, and further reports have not been found.

Liou and Wang described *U. coronatus* as originally observed on rice. There, the notation of black, obovate, oblong, or ellipsoidal teliospores, rounded and coronate at the apex, provides the basis for distinguishing this fungus from other species of *Uromyces*.

At present, rust diseases are not important on rice. Because of the well-known devastation of rust diseases on other cereals, they are included here to alert pathologists and breeders to their potential for damage in rice production as cultivars and cultural practices evolve.

Selected References

Diehl, W. W. 1944. Bibliography and nomenclature of *Puccinia oryzae*. Phytopathology 34:441-442.

Liou, T. N., and Wang, Y. C. 1935. Materials for study of rusts of China. V. Cont. Inst. Bot. Natl. Acad. Peiping 3:433-451.

Oh, S. H. 1985. Rice Diseases. 2nd ed. Commonwealth Mycological Institute, Kew, England.

(Prepared by R. K. Webster)

Leaf Sheath and Culm Diseases

Stem Rot

Stem rot is a serious disease that occurs in most rice-growing regions of the world. Estimates of yield loss due to the disease range from 10 to 75% (Plate 46).

Symptoms

The first symptoms are generally observed in the field after the mid tillering stage. Initially, the disease appears as irregular, black lesions on the outer leaf sheaths at the waterline (Plate 47). As the disease progresses, the lesions expand, and the fungus moves inward, forming lesions on the inner leaf sheaths (Plate 48). Eventually, the fungus penetrates and rots the culm (Plate 49). Infection of the culm can result in lodging, unfilled panicles, chalky grain, and in severe cases, death of the tiller. Numerous tiny, black sclerotia are embedded in diseased leaf sheath tissue (Plate 50), and sclerotia and mycelium of the fungus are generally present inside infected culms.

Causal Organism

Stem rot is caused by *Magnaporthe salvinii* (Cattaneo) R. Krause & R. K. Webster (syn. *Leptosphaeria salvinii* Cattaneo). The fungus is most commonly found in its sclerotial state, *Sclerotium oryzae* Cattaneo. The fungus also produces a conidial state, *Nakataea sigmoidea* (Cavara) K. Hara (syn. *Vakrabeeja sigmoidea* (Cavara) Subramanian). Perithecia of *M. salvinii* are found embedded in the leaf sheaths, and are dark, globose, and 250–650 μm in diameter, with a short beak (30–70 μm long) that generally does not protrude from the tissue. Asci are cylindrical, unitunicate, short-stalked, and 104–165 × 8.7–17.7 μm; they deliquesce at maturity and contain eight distichous ascospores. Ascospores are fusiform, somewhat curved, three-septate, and 35–65 × 8.7 μm. Sclerotia are black, globose or near globose, smooth, and usually 180–280 μm in diameter. Conidiophores are dark, upright, septate, and 100–175 × 4–5 μm. Conidia are fusiform, three-septate, curved, 29–49 × 10–14 μm, and produced on pointed sterigmata.

Disease Cycle and Epidemiology

The fungus survives between crops as sclerotia in crop debris or in soil. After flooding, sclerotia float to the surface of the paddy water, where they infect rice sheaths at the water line. Sclerotia are produced abundantly on diseased tissue as the disease progresses and the rice approaches maturity. The fungus also produces conidia and ascospores on infected plants that may serve as additional inoculum.

Disease incidence and severity are positively correlated with the number of sclerotia present in the upper layer of soil in a field before planting. The severity of disease increases with the amount of nitrogenous fertilizer and also depends on the time of infection. Field observations and greenhouse inoculation studies have shown that plants are most susceptible at the internode elongation stage and that infection at this stage results in the greatest reduction in yield. Isolates of *S. oryzae* have been shown to differ in virulence and severity of disease caused.

Control

Control of stem rot can be achieved through an integrated program of residue management, proper fertilization, and cultivar selection. Because most of the sclerotial inoculum is associated with crop residues, practices that reduce crop debris, such as burning, removal of straw, or hastening straw decomposition, lower the population of sclerotia. Deep burial of crop residues in the soil can also reduce inoculum by decreasing the availability and viability of sclerotia. Shallow incorporation of infested residues, however, may favor survival of inoculum. The use of nitrogenous fertilizers in excess of that needed to obtain maximum yields also favors severity of stem rot. Although no cultivars are completely resistant to the fungus, they do differ in susceptibility. Some show a high level of resistance, which has been shown to be quantitatively inherited. Sources of tolerance to stem rot have been identified in entries of the wild *Oryza* species *O. rufipogon* Griff. and *O. nivara* Sharma & Shastry. Control of the disease with fungicides has been shown to be effective but has not been widely adopted.

Selected References

Bockus, W. W., Webster, R. K., Wick, C. M., and Jackson, L. F. 1979. Rice residue disposal influences overwintering inoculum level of *Sclerotium oryzae* and stem rot severity. Phytopathology 69:862-865.

Ferreira, S. A., and Webster, R. K. 1976. Evaluation of virulence in isolates of *Sclerotium oryzae*. Phytopathology 66:1151-1154.

Figoni, R. A., Rutger, J. N., and Webster, R. K. 1983. Evaluation of wild *Oryza* species for stem rot (*Sclerotium oryzae*) resistance. Plant Dis. 67:998-1000.

Keim, R., and Webster, R. K. 1974. Nitrogen fertilization and severity of stem rot of rice. Phytopathology 64:178-183.

Krause, R. A., and Webster, R. K. 1972. The morphology, taxonomy, and sexuality of the rice stem rot fungus, *Magnaporthe salvinii* (*Leptosphaeria salvinii*). Mycologia 64:103-114.

Krause, R. A., and Webster, R. K. 1973. Stem rot of rice in California. Phytopathology 63:518-523.

(Prepared by R. K. Webster)

Sheath Blight

Among the fungal diseases affecting rice, sheath blight is second in importance only to rice blast and often rivals it. First reported in Japan by Miyake in 1910, the disease was first noted by Yano in 1901. It was soon established in many Asian countries and has been referred to as oriental leaf and sheath blight and as brown-bordered leaf and sheath spot. The pathogen affects rice and other plants in numerous and diverse habitats, and its host range is so broad that one authority considered it relatively safe to assert that no plant species is immune. The pathogen infects many weed grass species as well as rice and several major legume crops (soybeans, common bean, and mung bean) used in rotation with rice. The disease occurs throughout rice production areas of the world, in both tropical and temperate climates, where it is most damaging in intense production systems. Widely adopted changes in rice cultivation that favor sheath blight development have helped to make the disease significant worldwide.

The pathogen is soilborne and endemic in most rice production areas, and rice sheath blight can develop into a major production-limiting disease in a short time. The pathogen can be spread rapidly through irrigation water and by movement of soil during land preparation and cultivation. In about 10 years (1970–1980), sheath blight became the most important rice disease in the southern rice-producing areas of the United States. Yield losses can reach 50% with susceptible cultivars when the infection is well distributed and severe in a field.

Symptoms

Symptoms of sheath blight generally do not appear until plants are in the late tillering or early internode elongation growth stages. Occasionally symptoms are found on young plants. Initial symptoms consist of circular, oblong, or ellipsoid, green-gray, water-soaked spots about 1 cm long that occur on the leaf sheaths near the waterline (Plate 51). The lesions enlarge to approximately 2–3 cm in length and 1 cm in width, and the centers of the lesions become pale green or white and are surrounded by an irregular purple-brown border. Under favorable conditions, i.e., low sunlight, humidity near 95%, and high temperature (28–32°C), infection spreads rapidly by means of runner hyphae to upper plant parts, including leaf blades, and to adjacent plants (Plate 52). Lesions on the upper portion of plants may coalesce to encompass entire leaf sheaths and stems. Sclerotia are produced superficially on or near infected tissue after symptoms first appear (about 6 days) and are loosely attached and easily dislodged from the plant when they are mature.

Heavily infected plants produce poorly filled grains, particularly in the lower portion of the panicle (Plates 53 and 54). Additional losses in yield result from increased lodging or reduced ratoon production as a result of death of the culm.

Causal Organism

The causal agent of sheath blight is *Rhizoctonia solani* Kühn, teleomorph *Thanatephorus cucumeris* (A. B. Frank) Donk. The fungus belongs to anastomosis group AG-1, intraspecific group 1-A of *R. solani*.

Hymenia of *T. cucumeris* are effuse, cream to grayish white in color, and composed of a spidery network of repent hyphae from which loose patches of basidia arise in clusters. Basidia are barrel-shaped, slightly wider than the supporting hyphae, and average 14 (range 9–25) × 9 (5–12) μm. Sterigmata are long, stout, tapered, usually four (rarely two or three) per basidium, and average 13.5 (range 5 to 25–45) × 3 (to 4) μm. Basidiospores are hyaline, oblong to ellipsoid, often with one side flattened, with a prominent truncate apiculus, usually 9 (range 5–14.5) × 5.5 (4–8) μm, and capable of repetitive germination.

The mycelium of *R. solani* is colorless in young cultures and yellow-brown in older cultures. The hyphae have a prominent dolipore septum, are multinucleate, and are generally 8–12 μm in diameter. The young hyphae typically branch at 45 or 90° angles; hyphal branches are constricted at the point of origin and septate near it.

Three types of specialized mycelium produced by the pathogen are runner hyphae, lobate hyphae, and monilioid cells. The runner hyphae have thick, parallel walls and spread rapidly over the sheath and leaf surfaces of the rice plant. On the host, runner hyphae give rise at intervals to swollen, lobate hyphae (appressoria) or masses of appressoria (infection cushions) that produce infection pegs. Monilioid cells are short, broad cells produced in short chains that continue to form sclerotia. Sclerotia are irregularly hemispherical, flattened on the bottom, white when first formed and then turning brown or dark brown. On the host, they are formed on the surfaces of sheaths and leaves. Individual sclerotia measure up to 6 mm or more, but in culture they may coalesce to form larger masses. Initially, sclerotia are dense and will sink in water, but at maturity the cells in the outer layers become vacuolated, and the sclerotia become buoyant.

Disease Cycle and Epidemiology

The areas in a field affected initially may coalesce to form larger areas of an acre or more in extent. Disease develops most rapidly during early heading and grain filling. The pathogen survives between crops as soilborne sclerotia and mycelium in plant debris; these constitute the primary inoculum. Correlations between inoculum density and disease severity have been demonstrated. Sclerotia floating in the paddy tend to accumulate around the rice plant at the water-plant interface, so initial infections occur near the waterline. Sclerotia and sclerotial fragments can initiate hyphal growth several times. Sclerotia survive up to 2 years in temperate rice production areas and accumulate in the soil over time. Basidiospores of *T. cucumeris* can initiate infection but are considered to be relatively unimportant in the epidemiology of rice sheath blight.

The widespread adoption of new susceptible semidwarf cultivars and the changes in cultural practices associated with these cultivars favor development of sheath blight and have contributed greatly to its rapid increase in incidence and severity in rice-producing areas throughout the world. Because of their high yields, the shorter, more compact semidwarf cultivars require higher soil fertility, particularly in nitrogen. These cultivars are often direct seeded and grown in denser stands, factors that create a microclimate with high humidity and cooler temperatures and enhance disease development earlier in the growing season, thus making the disease more severe.

Control

Resistance to sheath blight has been investigated by researchers in several countries. No immunity has been found, but several cultivars with moderate to high resistance have been identified by the amount of disease spread upward on the plant. Resistance appears to be dominant or partially dominant. Tests of related wild rices have failed to provide sources more resistant than those found in cultivated species. Variation in resistance to sheath blight was observed in rice somaclones regenerated from the susceptible U.S. long-grain cultivar Labelle. Resistance of a somaclone with good agronomic characteristics was comparable to the most resistant cultivars previously identified and was stable after 5 years of field testing. Resistance was controlled by a single recessive gene in one somoclonal line and by two independently inherited recessive genes in another. Increased thickness of epicuticular wax appears to be correlated with resistance and with decreased production of lobate appressoria and infection cushions.

Preliminary investigations in Japan suggest that the Indica types of rice are more tolerant than the Japonica types. Among cultivars grown in the southern United States, the highest level of resistance is found in the short- and medium-grain types, which are more like the Japonica rices. Long-grain cultivars similar to the Indica rices possess little sheath blight resistance. Medium-grain cultivars in the United States suffer losses from 7 to 15%, whereas under the same conditions, long-grain cultivars suffer losses as high as 50% of potential yield. Grain weight, panicle length, and milling quality are affected by sheath blight.

Cultivar performance is greatly influenced by physiological, morphological, and ecological factors. Young plants are more resistant to sheath blight than older plants, with the greatest susceptibility occurring as panicles emerge. High levels of foliar nitrogen resulting from the use of large amounts of nitrogenous fertilizers are correlated with increased disease susceptibility. Short cultivars with broad leaves and high tillering abilities also produce a closed canopy (a favorable microclimate for sheath blight development) much earlier than tall cultivars with fewer tillers. Cultivars that mature early appear more susceptible than those that mature later, when environmental conditions may be less favorable for sheath blight development.

Foliar applications of fungicides to control sheath blight have been used commercially for many years in Japan. In the United States, foliar fungicides have been used since 1976, and certain fungicides have been found to give excellent control of sheath blight. Two fungicide applications are presently recommended for control of sheath blight, and the timing of these applications is very important. The first application should be made during the growth stage between early internode elongation and development of the 2.5- to 5-cm panicle in the boot. A second application should be made on 80–90% of emerging panicles from 10 to 14 days later. Several new fungicides specific for sheath blight control are being registered.

Classic approaches to inoculum manipulation or destruction for control of sheath blight are either not very effective or not feasible. Burning rice stubble to reduce sclerotial inoculum is often not feasible, and, if done, sometimes does not decrease sclerotial populations to levels that significantly reduce disease in the following crop. Long rotations with nonhost crops are generally not economically feasible. The use of rotation for disease control is also difficult, because the pathogen has a wide host range and can maintain its population on weeds and alternate hosts, such as soybeans and sorghum, that are commonly used in rotation with rice. Aerial blight of soybean, caused by the same pathogen, is increasing in incidence and severity in the United States. Prolific production of sclerotia on soybean also maintains high levels of inoculum in the soil.

Selected References

Anderson, N. S. 1982. The genetics and pathology of *Rhizoctonia solani*. Annu. Rev. Phytopathol. 20:329-347.

Groth, D. E., Rush, M. C., and Lindberg, G. D. 1990. Foliar fungicides for control of rice diseases in the United States. Pages 31-52 in: Pest Management in Rice. B. T. Grayson, M. B. Green, and L. G. Copping, eds. Elsevier Applied Science, London.

Jones, R. K., and Belmar, S. B. 1989. Characterization and pathogenicity of *Rhizoctonia* spp. isolated from rice, soybean, and other crops grown in rotation with rice in Texas. Plant Dis. 73:1004-1010.

Lee, F. N. 1980. Number, viability, and buoyancy of *Rhizoctonia solani* sclerotia in Arkansas rice fields. Plant Dis. 64:298-300.

Marshall, D. S., and Rush, M. C., 1980. Relation between infection by *Rhizoctonia solani* and *R. oryzae* and disease severity in rice. Phytopathology 70:941-946.

Marshall, D. S. and Rush, M. C. 1980. Infection cushion formation on rice sheaths by *Rhizoctonia solani*. Phytopathology 70:947-950.

Miyagi, Y. 1990. Fungicide use for the control of major rice diseases in Japan. Pages 111-121 in: Pest Management in Rice. B. T. Grayson, M. B. Green, and L. G. Copping, eds. Elsevier Applied Science, London.

O'Neill, N. R., and Rush, M. C. 1982. Etiology of sorghum sheath blight and pathogen virulence on rice. Plant Dis. 66:1115-1118.

O'Neill, N. R., Rush, M. C., Horn, N. L., and Carver, R. B. 1977. Aerial blight of soybeans caused by *Rhizoctonia solani*. Plant Dis. Rep. 61:713-717.

Oniki, M., Ogoshi, A., and Araki, T. 1986. Development of the perfect state of *Rhizoctonia solani* Kuhn AG-1. Ann. Phytopathol. Soc. Jpn. 52:169-174.

Ou, S. H. 1985. Rice Diseases. 2nd ed. Commonwealth Mycological Institute, Kew, England.

Parmeter, J. R., Sherwood, R. T., Jr., and Platt, W. D. 1969. Anastomosis grouping among isolates of *Thanatephoris cucumeris*. Phytopathology 59:1270-1278.

Talbot, P. H. B. 1970. Taxonomy and nomenclature of the perfect state. Pages 20-31 in: *Rhizoctonia solani*, Biology and Pathology. J. R. Parmeter, Jr., ed. University of California Press, Berkeley.

Xie, Q. J., Rush, M. C., and Lindberg, G. D. 1990. Somaclonal variation for disease resistance in rice (*Oryzae sativa* L.). Pages 491-509 in: Pest Management in Rice. B. T. Grayson, M. B. Green, and L. G. Copping, eds. Elsevier Applied Science, London.

(Prepared by M. C. Rush and F. N. Lee)

Sheath Spot

Sheath spot, also known as bordered sheath spot, red sclerotial disease, and Rhizoctonia sheath spot, was first reported from the United States in 1938 and has since been reported from Japan, Taiwan, Thailand, Vietnam, Cambodia, the Philippines, Brazil, and Africa. It is a relatively minor disease compared with sheath blight, which it closely resembles.

Symptoms

The lesions are usually found on leaf sheaths midway up the tiller and sometimes on the leaf blades (Plate 55A). Lesions are oval, 0.5–2 cm long and 0.5–1 cm wide. Lesions may be pale green, cream, or white, with a broad, dark, reddish brown margin. The lesions are usually separate and do not coalesce to form the large areas of continuous infection typical of sheath blight. The pathogen may also attack and weaken the culm under the sheath lesion on very susceptible cultivars, and weakened culms may lodge or break. Lodging caused by sheath spot is atypical, because infection usually occurs midway up rather than near the base of the culm.

Causal Organism

The disease is caused by *Rhizoctonia oryzae* Ryker & Gooch. Mycelium of *R. oryzae* is like that of *Rhizoctonia solani* Kühn. Sclerotia of *R. oryzae*, in contrast to those of *R. solani*, are generally amorphous and are salmon-colored to bright orange (Plate 55B). Sclerotia produced in diseased leaf sheaths are cylindrical. In culture, the sclerotia are submerged and usually form along main hyphal branches in a crow's-feet pattern. The teleomorph of *R. oryzae* closely resembles *Waitea circinata* Warcup & Talbot. Hymenia on infected sheaths are pink, resupinate, waxy-pruinose, and are easily detached from the leaf sheath. Basidia are urniform or suburniform, averaging 6.8 × 27.5 μm, and have four stout, cornute sterigmata, averaging 2–3 × 8.4 μm. Basidiospores are smooth, hyaline, broadly ellipsoid, usually with one side flattened, with a prominent truncate apiculus, and average 5.7 × 9.5 μm. The perfect state of *Rhizoctonia zeae* Voorhees, which is commonly isolated from rice sheaths with lesions similar to those of sheath spot, also conforms to *W. circinata*. Isolates of *R. oryzae* and *R. zeae* do not anastomose, and *R. zeae* isolates are morphologically distinct from *R. oryzae*, having small round, reddish brown sclerotia 0.5–1 mm in diameter.

Disease Cycle and Epidemiology

The pathogen survives as sclerotia and mycelium in soil and crop debris. Infections originate primarily from hyphae produced from sclerotia, which are hydrophobic and float on the paddy water, or from mycelium in floating plant debris. The hyphae grow up the rice tiller from the waterline and usually infect the leaf sheaths just below the leaf collar, several inches above the water. The perfect state is produced on lesions or on adjacent healthy tissue in the field during the heading stage, particularly under rainy, overcast conditions. Basidiospores can also cause infections but have not been demonstrated to be an important factor in disease incidence and spread. Single lesions produced in the middle and upper part of the rice canopy away from the water are attributed to basidiospore infection.

Control

The disease is of minor importance, and usually no attempts are made to control it. Cultivars with resistance to sheath blight also appear to be resistant to sheath spot. Only cultivars very susceptible to sheath blight are affected by the sheath spot disease. Certain fungicides used for sheath blight control also appear to give control of sheath spot.

Selected References

Gunnell, P. S., and Webster, R. K. 1984. The teleomorph of *Rhizoctonia oryzae*. (Abstr.) Phytopathology 74:857.

Hashioka, Y., and Makino, M. 1969. *Rhizoctonia* group causing the rice sheath spots in the temperate and tropical regions, with special reference to *Pellicularia sasakii* and *Rhizoctonia oryzae*. Res. Bull. Fac. Agric. Gifu Univ. 28:51-63.

Inagaki, K., and Adachi, T. 1987. Morphological and physiological characteristics of *Rhizoctonia* and *Sclerotium* species causing rice sclerotial diseases. (In Japanese; English summary.) Sci. Rep. Fac. Agric. Meijo Univ. 23:23-31.

Inagaki, K., Tamura, M., and Makino, M. 1987. Overwintering in plant residues of rice sclerotial disease fungi, *Rhizoctonia* and *Sclerotium* species. (In Japanese; English summary.) Sci. Rep. Fac. Agric. Meijo Univ. 29:27-29.

Jones, R. K., and Belmar, S. B. 1989. Characterization and pathogenicity of *Rhizoctonia* spp. isolated from rice, soybean, and other crops grown in rotation with rice in Texas. Plant Dis. 73:1004-1010.

Kuniyasu, K. 1983. The present situation of seedborne diseases and seed disinfection in Japan. Jpn. Pestic. Inf. 43:13-22.

Marshall, D. S., and Rush, M. C. 1980. Relation between infection by *Rhizoctonia solani* and *R. oryzae* and disease severity in rice. Phytopathology 70:941-946.

Ohata, K., Umehara, Y., and Ibaragi, T., eds. 1978. Diseases of Rice Seedlings Grown in Boxes and Methods for Preventing These Diseases. Nasangyoson-bunkakyokai, Takeda Chemical Company, Tokyo.

Oniki, M., Ogoshi, A., Araki, T., Sakai, R., and Tanaka, S. 1985. The perfect state of *Rhizoctonia oryzae* and *R. zeae*, and the anastomosis groups of *Waitea circinata*. Trans. Mycol. Soc. Jpn. 26:189-198.

Ou, S. H. 1985. Rice Diseases. 2nd ed. Commonwealth Mycological Institute, Kew, England.

Ryker, T. C., and Gooch, F. S. 1938. Rhizoctonia sheath spot of rice. Phytopathology 28:233-246.

(Prepared by M. C. Rush)

Aggregate Sheath Spot

Aggregate sheath spot, also known as the brown sclerotial disease, occurs in Japan, China, Vietnam, Thailand, India, and California.

Symptoms

The disease is characterized by oval lesions with a gray-green to straw-colored center surrounded by a distinct brown margin. A strip of necrotic cells runs down the middle and is conspicuous in young lesions. Often the lesions expand, forming a series of concentric bands. Initially, lesions occur on the lower leaf sheaths at the waterline and range from about 0.5 to 4 cm in length. As the disease progresses, lesions spread vertically up the rice plant to the upper leaf sheaths (Plate 56). Leaves of diseased sheaths generally turn yellow and die. The fungus may at times also infect the base of the leaf blade. Occasionally, the fungus causes a culm rot or infects the panicle rachis, resulting in sterile grains or partially filled heads (Plate 57). Aggregate sheath spot symptoms are similar to those of sheath blight, caused by *Rhizoctonia solani* Kühn, and sheath spot, caused by *R. oryzae* Ryker & Gooch.

Causal Organism

Aggregate sheath spot is caused by *Rhizoctonia oryzae-sativae* (Sawada) Mordue. The fungus is binucleate and belongs to the anastomosis group AG-Bb for binucleate *Rhizoctonia* isolates. In culture, *R. oryzae-sativae* produces white to pale brown mycelium and brown, irregularly globose sclerotia up to 2 mm in diameter that are composed of near-globose hyphal cells. On the host, the fungus produces globose sclerotia on the surface of infected rice sheaths as well as cylindrical sclerotia inside the cells of infected sheaths (Plate 58). The sexual state of the fungus is *Ceratobasidium oryzae-sativae* P. S. Gunnell & R. K. Webster; it occurs on the outside of leaf sheaths as an inconspicuous, effuse, whitish layer. The basidia are globose to subglobose, generally 13–21 × 12–18 μm, bearing two stout sterigmata, 12–23 × 4.5–5 μm. Basidiospores are globose to broadly ellipsoid, hyaline, repetitive, and 9–17 × 9–16 μm.

Disease Cycle and Epidemiology

The fungus survives between rice crops as sclerotia in soil or associated with infected debris. The fungus also infects some gramineous weeds, and these may serve as reservoirs of inoculum. In California, semidwarf cultivars have been found to be more susceptible to the disease than tall cultivars. Excess nitrogen does not favor the development of aggregate sheath spot, in contrast to sheath blight caused by *R. solani*.

Control

The disease is generally of minor importance, and control measures are not warranted. Using the least susceptible cultivars and practices that encourage degradation or entail removal of infected straw that harbors sclerotia minimize disease incidence.

Selected References

Gunnell, P. S., and Webster, R. K. 1984. Aggregate sheath spot of rice in California. Plant Dis. 68:529-531.

Gunnell, P. S., and Webster, R. K. 1987. *Ceratobasidium oryzae-sativae* sp. nov., the teleomorph of *Rhizoctonia oryzae-sativae* and *Ceratobasidium setariae* comb., nov., the probable teleomorph of *Rhizoctonia fumigata* comb. nov. Mycologia 79:731-736.

(Prepared by P. S. Gunnell)

Sheath Net Blotch

Sheath net blotch has only been reported from Japan and is a minor disease.

Symptoms

The fungus produces oval lesions, 1–2 cm long, which occur on leaf sheaths near the waterline. Lesions are initially small, yellow, water-soaked spots. Over time the lesions expand, become whitish, and are covered by a brown, netlike pattern. This characteristic pattern is formed from horizontal and longitudinal streaks. Infected leaves may yellow and die.

Causal Organism

Cylindrocladium scoparium Morg. produces erect, hyaline, conidiophores, up to 500 μm in length, which branch repeatedly near the top. The main axis of the conidiophore usually is elongate, ending in a terminal vesicle or swelling that is clavate and measures 19–22 × 7–11 μm. Phialides at the tips of the ultimate branches produce conidia that are hyaline, cylindrical with rounded ends, uniseptate, and 50–58 × 5–6 μm. The fungus also produces microsclerotia composed of groups of chlamydospores.

Disease Cycle and Epidemiology

C. scoparium can survive on diseased straw in the field. It has a broad host range, attacking other grasses, legumes, and curcurbits, and alternate hosts may serve as a source of primary inoculum.

Control

The disease is of minor importance, and control practices are generally not warranted.

Selected References

Fujikawa, T. 1954. Studies on the sheath net-blotch of rice plant. II. Survival and host plants of the pathogen. Kyushu Agric. Res. 3:107-108.

Fujikawa, T., Tomiku, T., and Okatome, Z. 1962. Varietal difference in resistance against sheath net-blotch in rice plants. Nogyo oyobi Engei 37:395-396.

Fujikawa, T., Tomiku, T., and Sato, S. 1970. Studies on the sheath net-blotch of rice plant. XII. Effects of various fungicides on the disease. Proc. Assoc. Plant Protect. Kyushu 16:87-89.

Nonaka, F., and Morita, A. 1973. Seasonal development of sheath net-blotch of rice plant and its causal organism, *Cylindrocladium scoparium* Morgan. Agric. Bull. Suga Univ. 34:29-33.

(Prepared by D. Groth)

Sheath Rot

Sheath rot has been reported from Japan, India, Bangladesh, Sri Lanka, Kenya, Nigeria, The Gambia, Mexico, and the United States. The disease has commonly caused yield losses of 20–85%.

Symptoms

Lesions occur on the upper leaf sheaths, especially the flag leaf sheath. Lesions are initially oblong, 5–15 mm in length, with gray to light brown centers surrounded by a dark reddish brown margin (Plate 59). As the disease progresses, lesions enlarge and coalesce and may cover most of the leaf sheath. Lesions may also consist of a diffuse reddish brown discoloration in the sheath. With early or severe infection, the panicle may fail to emerge completely or at all. Panicles that have not emerged tend to rot, and florets turn red-brown to dark brown (Plate 60). Sterile, shriveled, or partially filled grain is also associated with infection of the panicle.

Causal Organism

Sheath rot is caused by *Sarocladium oryzae* (Sawada) W. Gams & D. Hawksw. (syn. *Acrocylindrium oryzae* Sawada). *S. attenuatum* W. Gams & D. Hawksw. has also been reported to cause sheath rot of rice but is now regarded as a synonym of *S. oryzae*. The fungus produces verticillate conidiophores with one to two whorls of appressed branches. The ultimate branches are phialides and produce cylindrical to slightly fusiform, often somewhat curved, hyaline, single-celled conidia, 3.5–9 × 1–2.5 μm. Phialidic conidiogenous cells may also arise directly from the hyphae. Condiophores and conidia often form a white layer on the inside of infected sheaths.

Disease Cycle and Epidemiology

S. oryzae survives as mycelium in infected residue and on seeds. It also infects several weed hosts and bamboo. Isolates of the fungus from weeds (but not from bamboo) have been shown to infect rice. The fungus invades rice through the stomata and wounds and grows intercellularly in the vascular bundles and mesophyll tissues. Insect damage, especially from mites, stem borers, and mealy bugs, aids disease development by retarding panicle emergence and providing entry wounds for the fungus. Most infected plants exhibit some type of insect damage. The disease appears to be favored by low nitrogen levels or dense stands.

Control

Certain cultivars have resistance to the disease. Also, some strains of *Pseudomonas fluorescens* have been reported to suppress sheath rot.

Selected References

Amin, K. S. 1974. Sources of resistance to Acrocylindrium sheath-rot of rice. Plant Dis. Rep. 60:72-73.

Boa, E. R., and Brady, B. L. 1987. *Sarocladium oryzae* associated with blight of *Bambusa* species in Bangladesh. Trans. Br. Mycol. Soc. 89:161-166.

Bridge, P. D., Hawksworth, D. L., Kavishe, D. F., and Farnell, P. A. 1989. A revision of the species concept in *Sarocladium*, the causal agent of sheath-rot in rice and bamboo blight, based on biochemical and morphometric analyses. Plant Pathol. 38:239-245.

Chakravartz, D. K., and Biswas, S. 1978. Estimation of yield loss in rice affected by sheath rot. Plant Dis. Rep. 62:226-227.

Sakthivel, N., and Gnanamanickam, S. S. 1987. Evaluation of *Pseudomonas fluorescens* for suppression of sheath rot disease and for enhancement of grain yields in rice (*Oryza sativa* L.). Appl. Environ. Microbiol. 53:2056-2059.

(Prepared by D. Groth)

Crown Sheath Rot

Crown sheath rot, also known as black sheath rot, Arkansas foot rot, and brown sheath rot, has been reported from the United States, Japan, India, Australia, the Philippines, Africa, and India. It is usually a minor disease but occasionally causes serious damage to susceptible cultivars. Some reports of damage attributed to crown sheath rot may be suspect, because the symptoms can be confused with those of rice stem rot.

Symptoms

Symptoms usually become evident at the time of internode elongation. The disease affects the leaf sheaths, causing a dark brown to black discoloration or lesion that extends upward from the crown (Plate 61). Leaves of heavily infected sheaths die. The culm and crown are also infected, which results in inhibition of tillering and contributes to lodging, incomplete grain filling, and premature ripening of the panicle. Dark mycelial mats are visible on the inner surface of diseased sheaths and on the culm. Within infected sheaths, perithecia are formed, and their necks protrude through the sheath epidermis.

Causal Organism

Crown sheath rot is caused by *Gaeumannomyces graminis* (Sacc.) Arx & D. Olivier var. *graminis*. Perithecia of the fungus are black, immersed, globose to oval, 200–400 μm in diameter or 200–400 × 150–300 μm, with an erumpent perithecial neck, generally 100–500 × 7–100 μm. Asci are unitunicate, elongate, and clavate, have a distinct apical ring, and contain eight ascospores (Plate 62). Ascospores are filiform, tapered to the base, slightly curved, hyaline to somewhat yellowish, three- to five-septate, and 80–100 × 2.5–3 μm. Paraphyses are hyaline, septate, and 8–10 μm wide at the base, tapering to 2–3 μm at the apex. Hyphae are brown and septate. Hyphopodia are lobed and terminal, 15–25 (up to 30) × 10–20 (up to 25) μm or simple (unlobed) and circular to oval, 7–15 × 4–7 μm. Single-celled, hyaline conidia (4–7 × 1–1.5 μm) are produced from clusters of hyaline phialides (9–18 × 2–3 μm). The fungus differs from *G. graminis* var. *tritici* J. Walker in producing lobed hyphopodia.

Disease Cycle and Epidemiology

Information is scant. *G. graminis* var. *graminis* persists in infected rice debris and perhaps on alternate hosts. The fungus has been reported to infect *Typha latifolia* and species of *Pennisetum, Stenotaphrum, Triticum, Axonopus, Chloris,* and *Cynodon*. The fungus infects rice through the sheaths or basal leaves by direct penetration of mycelium or by penetration pegs formed from hyphopodia. Ascospores are forcibly ejected from the perithecia after periods of wetness and are likely disseminated by rain and/or wind.

Control

Cultivars vary in susceptibility. Those that are highly susceptible should not be grown.

Selected References

Ou, S. H. 1985. Rice Diseases. 2nd ed. Commonwealth Mycological Institute, Kew, England.

Padwick, G. W. 1950. Manual of Rice Diseases. Commonwealth Mycological Institute, Kew, England.

Tullis, E. C. 1933. *Ophiobolus oryzinus* the cause of a rice disease in Arkansas. J. Agric. Res. 46:799-806.

Tullis, E. C., and Adair, C. R. 1947. Black sheath rot of rice (*Ophiobolus oryzinus*) caused lodging of rice in Arkansas and Texas in 1947. Plant Dis. Rep. 31:468.

Walker, J. 1973. *Gaeumannomyces graminis* var. *graminis*. Descriptions of Pathogenic Fungi and Bacteria, No. 382. Commonwealth Mycological Institute, Kew, England.

(Prepared by D. Groth)

Myrothecium Blotch

Myrothecium blotch has been reported from Italy and India.

Symptoms

Irregular to oblong lesions, 4–20 cm in length, with dark brown centers and a lighter brown margin, occur on the flag leaf sheaths. Brown lesions, sometimes with whitish centers, also occur on the glumes of affected plants.

Causal Organism

The disease is caused by *Myrothecium verrucaria* (Albertini & Schwein) Ditmar:Fr. Sporodochia are at first green, then black, with a white margin of contorted hyphae, up to 1.5 mm in diameter, with no setae. Condiophores are phialides, 10–12 × 1–2 μm. Conidia are one-celled, ovoid to ellipsoid, pale olive, and 6–10 × 2–4.5 μm.

Disease Cycle and Epidemiology

Little information on epidemiology or control is available.

Selected References

De Carolis, C. 1972. *Myrothecium verrucaria* (Alb. et Schw.) Ditm. ex Fr. su riso in Italia. Riso 21:253-258.

Nguyen, T. N., Mathur, S. B., and Neergaard, P. 1973. Seed-borne species of *Myrothecium* and their pathogenic potential. Trans. Br. Mycol. Soc. 61:347-354.

(Prepared by P. S. Gunnell)

Sheath Blotch

Sheath blotch has been reported from Japan, Burma, China, India, Malaysia, Sri Lanka, Thailand, the Philippines, Sierra Leone, Swaziland, Fiji, and the Solomon Islands.

Symptoms

The disease usually occurs at the late tillering stage and is characterized initially by large, brown, oblong lesions, about 2.5-cm long, on the leaf sheaths. As the disease progresses, the lesions lengthen up to 10 cm, and turn turn grayish at the center . Affected sheaths eventually become desiccated.

Causal Organism

Sheath blotch is caused by *Pyrenochaeta oryzae* Shirai ex Miyake. Pycnidia are immersed, dark brown, subglobose, up to 220 μm in diameter, ostiolate, and with 5–20 dark brown setae surrounding the ostiole. Setae are dark brown, cylindrical, slightly tapered, and 50–140 × 4–5 μm. Conidiophores are septate, branched, hyaline, and phialidic. Conidia are hyaline, fusiform, biguttulate, and 4–6 × 1–1.5 μm.

Disease Cycle and Epidemiology

Spores of the fungus are presumably dispersed by splashing water. Lesions often appear at oviposition sites of the green leafhopper.

Control

No control measures have been investigated, because the disease is of minor importance.

Selected References

Ou, S. H. 1985. Rice Diseases. 2nd ed. Commonwealth Mycological Institute, Kew, England.

Punithalingam, E. 1980. *Pyrenochaeta oryzae*. Descriptions of Pathogenic Fungi and Bacteria, No. 666. Commonwealth Mycological Institute, Kew, England.

(Prepared by P. S. Gunnell)

Root and Crown Diseases

Bakanae

Bakanae, one of the first diseases of rice to be described scientifically, was recognized in Japan in 1828 and formally described in 1898. The disease is widely distributed in Asia and has been responsible for yield losses of up to 50%. *Bakanae* (pronounced *ba-ka-ni*) is a Japanese word meaning "foolish seedling" and refers to the abnormal elongation often seen in infected plants. Study of bakanae and the causal fungus led to the discovery of the plant growth hormone gibberellin.

Symptoms

The classic syndrome associated with bakanae is elongated, slender, pale seedlings that are conspicuous next to healthy seedlings (Plates 63 and 64). Infected seedlings may, however, also be stunted and chlorotic, exhibiting root and crown rot. Infected seedlings usually are killed. Older plants can also be infected and may exhibit abnormal elongation and produce adventitious roots at the lower nodes of the culm. Plants that survive until maturity are sterile and produce no panicles or empty panicles. The term *palay lalake* or "man rice" is used to describe the sterility phenomenon in the Philippines. Also, the leaf sheaths of infected plants may turn blue-black with the production of perithecia. The panicles of healthy plants may be infected and turn pink, and the hulls or seeds may have a reddish color due to growth and sporulation of the fungus. Pink panicles can be caused by other species of *Fusarium* as well.

The types of symptoms produced are determined by the strain of the fungus and the inoculum potential of the pathogen. The fungus produces the growth hormones gibberellin and fusaric acid. Gibberellin is responsible for plant elongation, whereas fusaric acid causes stunting. The relative amounts of gibberellin and fusaric acid produced by the fungus are dependent on the particular strain and nutritional conditions. High inoculum densities in soil reportedly result in seedling blight, stunting, and chlorosis, whereas moderate inoculum densities result in the classic bakanae symptoms. Heavily infested seeds have been variously reported to produce stunted seedlings or both stunted and elongated seedlings, whereas seedlings from lightly infested seed (not discolored) reportedly produce bakanae symptoms only.

Causal Organism

Gibberella fujikuroi (Sawada) Wollenweb. var. *fujikuroi* (anamorph *Fusarium moniliforme* J. Sheld.) causes bakanae of rice. Perithecia of the fungus are dark blue and globose to conical, 220–300 × 250–300 μm. Asci are ellipsoid to clavate and contain four to eight monostichous to distichous ascospores. Ascospores are hyaline, elliptical, with one to sometimes three septa, 10–20 × 4–7 μm. Microconidia are hyaline, oval to clavate, single-celled, 5–12 × 1.5–2.5 μm, and are formed in chains or false heads from monophialides. Macroconidia are somewhat sickle-shaped to almost straight, with a foot-shaped basal cell, thin walls, and usually three to five (but up to seven) septa, and measure 25–60 × 2.5–4 μm. The fungus does not produce chlamydospores. Dark blue sclerotia are sometimes present.

Disease Cycle and Epidemiology

Bakanae is seedborne and is primarily seed-transmitted. Seed infection occurs before harvest via airborne ascospores and also from conidia that contaminate the seed during harvesting. The fungus does not infest the seed internally. *G. fujikuroi* may also survive in the soil in infested crop residue as thick-walled hyphae or macroconidia, but it is relatively short-lived in soil. The fungus infects plants through the roots or crowns and grows systemically in the plant but does not systemically infect the panicle. The disease is favored by high temperatures (30–35°C).

Control

The use of clean seed is essential to control bakanae. Seed treatment with fungicides is effective. Salt water can be used to separate lightweight, infected seeds from seed lots, thereby reducing seedborne inoculum. Cultivars vary in resistance to the disease, and in areas where fungicide-treated seed is not available, avoiding the use of highly susceptible cultivars can reduce the disease incidence and severity.

Selected References

Kuhlman, E. G. 1982. Varieties of *Gibberella fujikuroi* with anamorphs in *Fusarium* section *Liseola*. Mycologia 74:759-768.

Ou, S. H. 1985. Rice Diseases. 2nd ed. Commonwealth Mycological Institute, Kew, England.

Sun, S. K. 1975. The disease cycle of rice bakanae disease in Taiwan. Proc. Natl. Sci. Counc. Repub. China 8(2):245-256.

Sun, S. K., and Snyder, W. C. 1981. The bakanae disease of the rice plant. Pages 104-113 in: *Fusarium*: Diseases, Biology, and Taxonomy. P. E. Nelson, T. A. Toussoun, and R. J. Cook, eds. Pennsylvania State University Press, University Park.

Sung, J. M., Yang, S. S., Lee, E. J., and Park, J. S. 1983. Classification of *Fusarium moniliforme* isolates into four strains based on mycological characteristics. Korean J. Mycol. 11(4):169-175.

Yu, K. S., and Sun, S. K. 1977. Studies on inoculum potential and incubation period of rice bakanae disease. (In Chinese; English summary.) Plant Prot. Bull. 19:245-250.

(Prepared by J. M. Bonman)

Feeder Root Necrosis and Root Rot

Feeder root necrosis (FRN) is a recently coined descriptive term for a disorder that has caused decline problems on several crop plants. FRN has been referred to as a cryptic disease, because there are no readily apparent symptoms.

FRN of rice has been shown to be caused solely by *Pythium* species. A separate disease, rice root rot, which also has been associated with *Pythium* spp., has been shown to be caused by an interaction between *Pythium* spp. and the rice water weevil. Unlike FRN, rice root rot has defined symptoms, characterized by a browning and necrosis of the roots, failure to respond to fertilizers, uneven maturity, and reduced grain yields.

Symptoms

FRN is difficult to diagnose. Infected plants do not wilt or collapse, and there is no obvious rot of the primary roots. Root symptoms are characterized by an overall lack of fine feeder roots, and there may be restricted, reddish brown lesions on the primary roots where the feeder roots were attached (Plates 65 and 66). In extreme cases, plants may be slightly stunted and appear unthrifty.

Causal Organisms

Pythium spinosum Sawada, *P. dissotocum* Drechs., *P. irregulare* Buisman emend. Vaartaja, *P. arrhenomanes*

Drechs., *P. myriotylum* Drechs., *P. catenulatum* Matthews, and *P. graminicola* Subramanian have been isolated from rice roots and have been found to be pathogenic. Strains within a species have been found to vary greatly in virulence.

Disease Cycle and Epidemiology

Little is known about the disease cycle and epidemiology of FRN, because it has been described only recently. However, the disease may be widespread, since the *Pythium* species associated with FRN have broad host ranges and are capable of surviving under diverse climatic conditions. Preliminary experimental results suggest that FRN is favored by cool temperatures and indicate that the disease is not a problem in drill-seeded rice. Feeding wounds caused by larvae of the rice water weevil, *Lissorhoptrus oryzaophilus* Kuschel, predispose the rice roots to infection by *Pythium* spp.

Control

Excellent control of FRN has been obtained with the use of systemic fungicides specific for *Pythium* spp. applied to the soil immediately prior to flooding and seeding.

Selected References

Atkins, J. G. 1974. Rice Disease of the Americas: A Review of the Literature. U.S. Dep. Agric. Agric. Handb. 448.

Hendrix, F. F., Jr., and Campbell, W. A. 1983. Some pythiaceous fungi: New roles for old organisms. Pages 123-160 in: Zoosporic Plant Pathogens: A Modern Perspective. S. T. Buczacki, ed. Academic Press, New York.

Salt, G. A. 1979. The increasing interest in "minor pathogens." Pages 289-312 in: Soilborne Plant Pathogens. B. Schippers and W. Gams, eds. Academic Press, New York.

Schneider, R. W., Rush, M. C., and Pillay, M. 1988. Feeder root necrosis recognized as a serious disease in rice. La. Agric. 31:5-7.

(Prepared by R. Schneider)

Grain Diseases

False Smut

False smut, also known as green smut, is found throughout tropical Asia and in Italy, Australia, Costa Rica, the Dominican Republic, Panama, Mexico, South America, and the United States. False smut is generally a minor disease, but epidemics of the disease have been reported in India, Burma, Peru, and the Philippines.

Symptoms

The rice kernels are replaced by globose, velvety spore balls up to 1 cm in diameter (occasionally up to 5 cm), which burst out from between the glumes (Plate 67). The balls consist of three spore-producing layers surrounding a hard core of tightly woven mycelium. The innermost and middle layers contain immature spores and are pale yellow and orange yellow, respectively; and the outer layer, which consists of mature spores, is olive to olivaceous-black. Immature spore balls appear orange and are covered by a membrane. In most cases, the ovary is destroyed, whereas the style, stigma, and anthers remain intact and are incorporated into the spore ball. At the center of the spore ball, one or more irregular, hard, black sclerotia, 5–13 × 2–5 mm, develops.

In general, only a few grains of a panicle are affected. Some researchers have reported that the grains adjacent to smut balls are sterile.

Causal Organism

False smut is caused by *Ustilaginoidea virens* (Cooke) Takah. Conidia of the fungus, sometimes referred to as chlamydospores, are spherical to elliptical, pale and almost smooth when young, olivaceous and warty when mature, and 3–5 × 4–6 μm; they are formed along the hyphae on tiny sterigmata. In culture, the conidia reportedly germinate by germ tubes that form small, ovoid secondary conidia. Stipitate stromata with globose heads grow from the sclerotia, and at the periphery, the heads are embedded with perithecia. The perithecia contain cylindrical, eight-spored asci, 180–220 × 4 μm. Ascospores are filiform, hyaline, nonseptate, 120–180 × 1.5 μm (50–80 μm according to Hashioka). The sexual state was named *Claviceps oryzae-sativae* by Hashioka, but he incorrectly described the stromata that emerged from the sclerotia as apothecia.

Disease Cycle and Epidemiology

The disease cycle of false smut has not been completely characterized. The fungus usually invades the ovary at the early flowering stage, but it can also invade mature grain. Conidia, secondary conidia, and ascospores are all reportedly capable of infection. The fungus can survive as sclerotia or hardened spore balls, called pseudomorphs. Pseudomorphs have been reported to survive up to 4 months under field conditions. *U. virens* also can infect *Panicum trypheron* Schultes., *Digitaria adscendens* (H. B. K.) Henrard (syn. *D. marginata* Link), and wild *Oryza* spp., which may serve as important reservoirs of inoculum between rice crops in tropical areas. Periods of rain, high humidity, and soils high in nitrogen are reported to favor disease development.

Control

No control measures are generally warranted for false smut. Fungicides applied at heading have been reported to be successful in controlling the disease, and some cultivars have been reported to be resistant to the fungus.

Selected References

Gangopadhay, S. 1983. Current Concepts on Fungal Diseases of Rice. Today and Tomorrow's Printers & Publishers, New Delhi, India.

Hashioka, Y. 1971. Rice diseases in the world—VIII. Diseases due to Hypocreales, Ascomycetes (fungal diseases—5). Riso 20:235-258.

Mulder, J. L., and Holliday, P. 1971. *Ustilaginoidea virens*. Descriptions of Plant Pathogenic Fungi and Bacteria, No. 299. Commonwealth Mycological Institute, Kew, England.

Ou, S. H. 1985. Rice Diseases. 2nd ed. Commonwealth Mycological Institute, Kew Surrey, England.

Sudheer, A., and Shetty, H. Shekara. 1987. Role of *Panicum trypheron* in annual recurrence of false smut of rice. Trans. Br. Mycol. Soc. 88:409-411.

(Prepared by F. N. Lee and P. S. Gunnell)

Kernel Smut

Kernel smut was described independently in Japan in 1896 and in the United States in 1899. Besides Japan and the United States, the disease has been reported in India, Burma, China, the Philippines, Taiwan, Indonesia, Thailand, Vietnam, Malaysia, Nepal, Pakistan, Mexico, Guyana, Trinidad, Brazil, Venezuela, Panama, and Sierra Leone.

Kernel smut causes losses in both yield and quality, and yield losses as high as 15% have been reported.

Color Plates

1. Seedlings grown in nursery boxes for mechanical transplanting.

2. Chlorosis of the second and third leaves of rice seedlings caused by *Pseudomonas plantarii*. (Courtesy K. Azegami)

3. Seedlings with symptoms of seedling blight in a nursery box. (Courtesy K. Azegami)

4. Chlorotic patches of seedlings infected with seedling blight. (Courtesy K. Azegami)

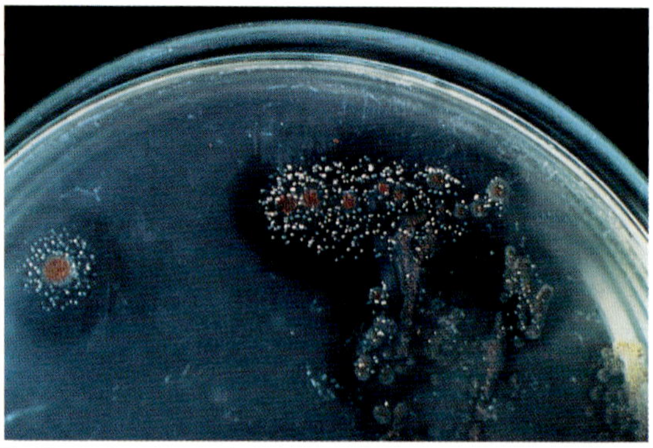

5. Tropolone produced by *Pseudomonas plantarii* visible as a darkened area around the bacterial colonies. (Courtesy K. Azegami)

6. Bacterial brown stripe of seedlings caused by *Pseudomonas syringae* pv. *panici*.

7. Symptoms of bacterial brown stripe in a wet bed nursery.

8. Field symptoms of sheath brown rot on the flag leaf sheath. (Courtesy K. Miyajima)

9. Close-up of sheath brown rot lesions. (Courtesy K. Miyajima)

10. Seedling rot in the field after transplanting, caused by *Pseudomonas glumae*. (Courtesy K. Miyajima)

11. Panicles infected with grain rot caused by *Pseudomonas glumae*.

A B

12A and B. Panicles infected with palea browning caused by *Erwinia herbicola*. (Courtesy K. Azegami)

13. A field showing severe bacterial leaf blight infection at the maximum tillering stage.

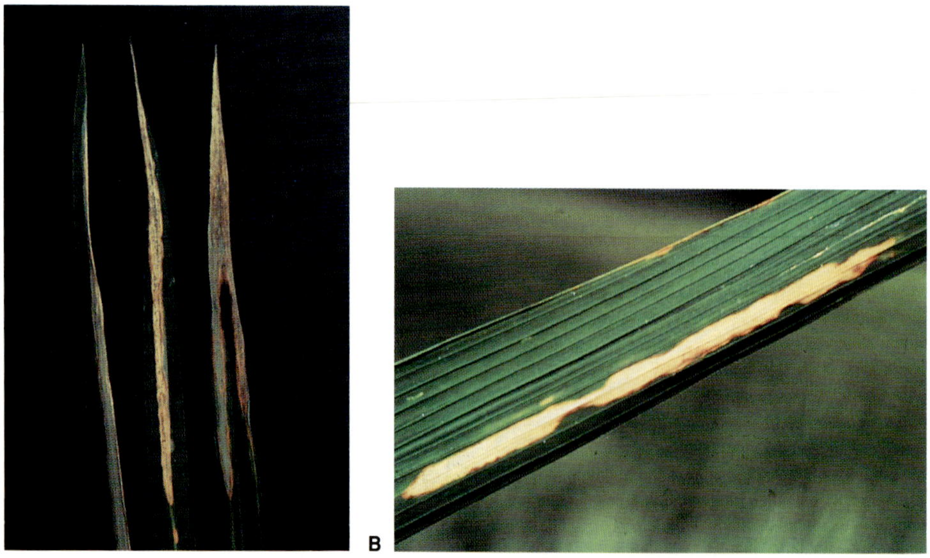

A B

14. Bacterial leaf blight, caused by *Xanthomonas oryzae* pv. *oryzae*, begins as water-soaked stripes on leaf blades (**A**); the stripes enlarge, becoming yellow and then white (**B**).

15. Kresek syndrome of bacterial leaf blight.

16. Pale yellow leaf syndrome of bacterial leaf blight.

17. Plants exhibiting the kresek syndrome inoculated with *Xanthomonas oryzae* pv. *oryzae* at 9, 16, 23, and 32 days (from left to right).

18. Plants in the field infected with bacterial leaf streak.

19. Foliar symptoms of bacterial leaf streak showing typical water-soaked lesions.

20. Drops of ooze of *Xanthomonas oryzae* pv. *oryzicola* from bacterial leaf streak lesions.

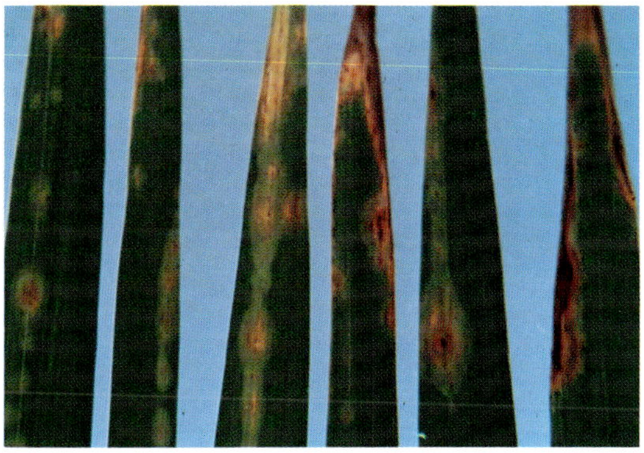

21. Halo blight lesions caused by *Pseudomonas syringae* pv. *oryzae*. (Courtesy H. Kuwata)

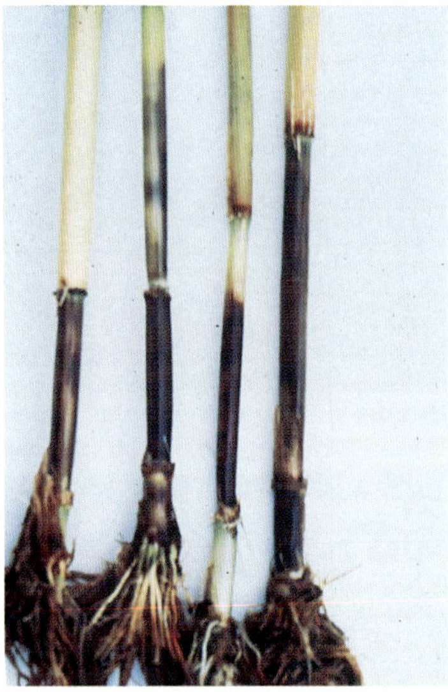

22. Tillers infected with foot rot, showing blackening of the internodes and root rot.

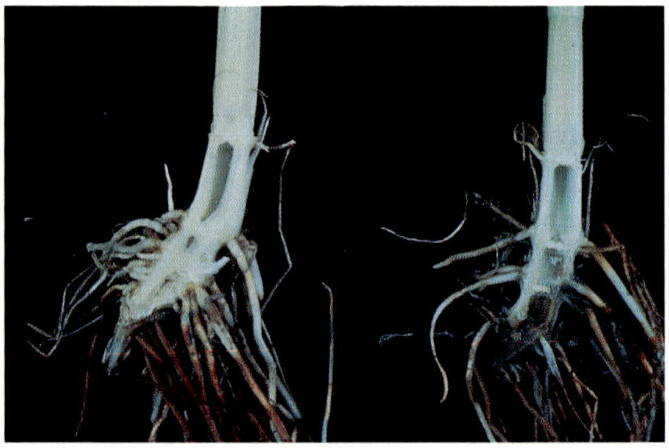

23. Longitudinal section through a culm infected with foot rot, showing internal ooze of *Erwinia chrysanthemi*.

24. Yellowing and wilting of leaves caused by a severe foot rot infection.

25. Water mold disease of water-seeded rice. Hyphal strands typical of *Achlya* sp. on rice. (Courtesy R. K. Webster)

26. Water mold of water-seeded rice in a field that was drained after sowing.

27. Seedling blight lesions on rice.

28. Seedbeds for producing rice transplants.

29. Seedbeds affected by seedling blight, caused by *Sclerotium rolfsii*.

30. Young blast lesions caused by *Pyricularia oryzae*.

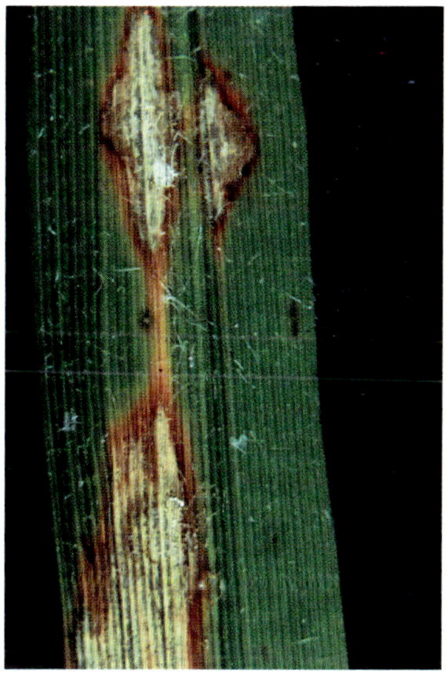

31. Mature blast lesions showing necrotic borders.

32. Sporulating blast lesion.

33. Foliar blast symptoms.

34. Collar blast.

35. Infection of the panicle node, or neck blast, compared with healthy node.

36. Panicle with neck blast.

A

B

37A and B. Brown spot leaf lesions. (Courtesy D. Groth)

38. Leaf scald lesions usually begin at the leaf tips.

39. Lesions of leaf scald may coalesce as they mature.

40. Narrow brown spot of rice, caused by *Cercospora oryzae*. (Courtesy D. Groth)

41. Stackburn, caused by *Alternaria padwickii*.

42. Leaves infected with leaf smut show small spots or streaks produced by sori of *Entyloma oryzae*.

43. Eyespot of rice.

44. Field view of severe eyespot.

45. Twisted grain and grain abortion caused by downy mildew.

46. Severe infection of stem rot in field.

47. Initial symptoms of stem rot on rice.

48. Advanced infection of stem rot.

49. Severe infection of culms by stem rot.

50. Sclerotia of *Sclerotium oryzae* in leaf sheath.

51. Initial sheath blight lesion following infection by *Rhizoctonia solani.* (Courtesy M. C. Rush)

52. Severe sheath blight lesions on rice leaves and sheaths. (Courtesy M. C. Rush)

53. Severe sheath blight in the field. (Courtesy M. C. Rush)

54. Sheath blight on the panicle. (Courtesy D. Groth)

A

B

55. A, Sheath spot, caused by *Rhizoctonia oryzae*. **B,** Hymenium with sclerotia of *R. oryzae*. (B, courtesy P. S. Gunnell)

56. Aggregate sheath spot lesions on leaves, sheaths, and culm.

57. Aggregate sheath spot lesions on flag leaves and panicles.

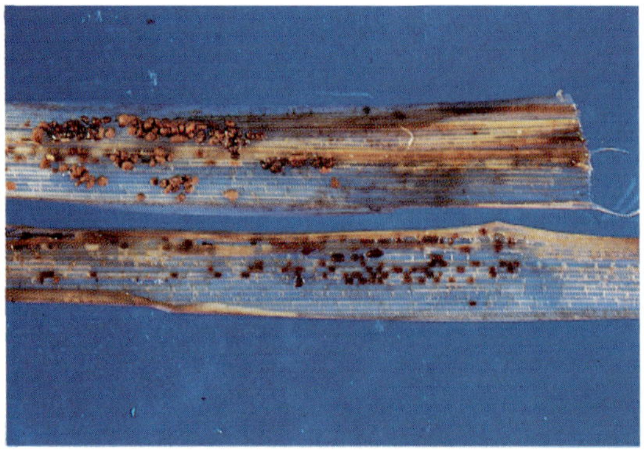

58. Sclerotia of *Rhizoctonia oryzae-sativae* in leaf sheaths.

59. Sheath rot symptoms on the flag leaf sheath.

60. Sheath rot on panicle and grain.

61. Crown sheath rot symptoms with perithecia of *Gaeumannomyces graminis* var. *graminis* visible. (Courtesy F. N. Lee)

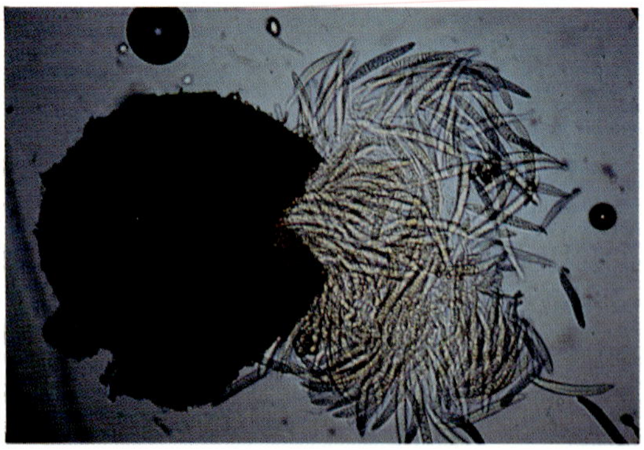

62. Perithecia and ascospores of *Gaeumannomyces graminis* var. *graminis*. (Courtesy F. N. Lee)

63. Elongated, pale seedlings typical of bakanae disease next to healthy seedlings in the field.

64. Seedlings affected with bakanae disease (right) and healthy seedlings (left).

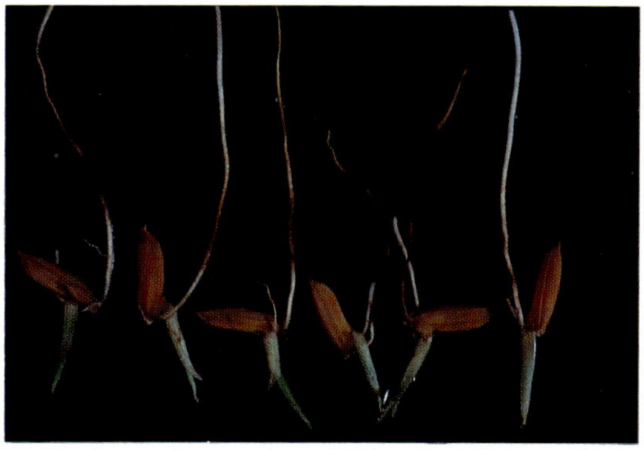

65. Feeder root necrosis on field-grown rice seedlings.

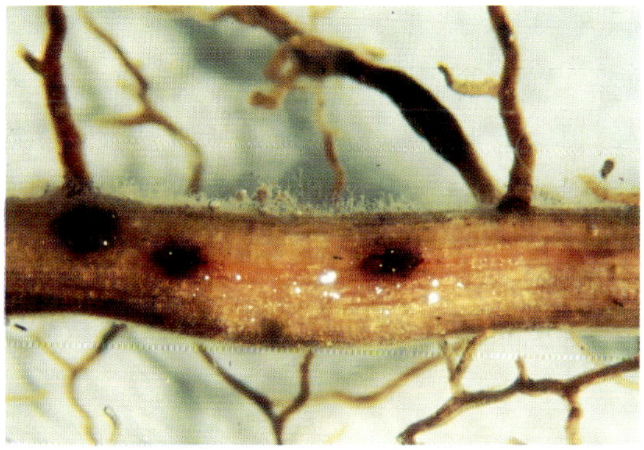

66. Localized lesions on the primary root, caused by spread of feeder root necrosis infection.

67. Velvety spore balls of false smut. (Courtesy M. C. Rush)

68. Kernel smut.

69. Udbatta disease. Note erect panicle covered by mycelium. (Courtesy F. N. Lee)

70. Field symptoms of glume blight. (Courtesy A. S. Prabhu)

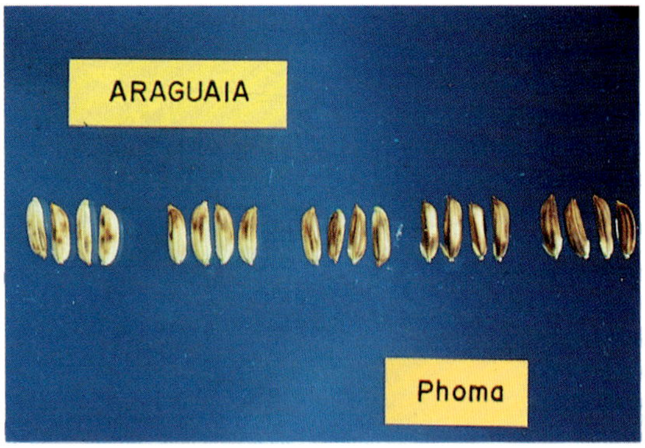

71. Varied degrees of glume discoloration on kernels (cv. Araguaia) inoculated with *Phoma sorghina* at the boot stage. (Courtesy A. S. Prabhu)

72. Scab of the panicle and grain.

73. Grain discoloration typical of insect and micro-organism attack of mature grain.

74. Ear blight.

75. Stunting and dark green leaves caused by black-streaked dwarf virus.

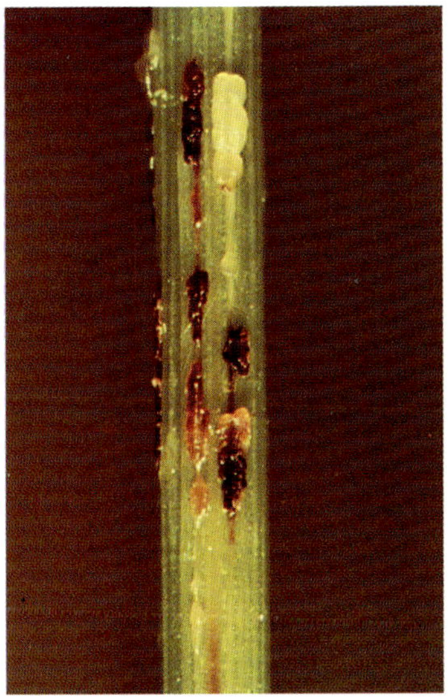

76. Galls on a rice culm caused by black-streaked dwarf virus. (Courtesy A. Shinkai)

77. A female small brown planthopper, *Laodelphax striatellus*.

78. A female green leafhopper, *Nephotettix virescens*.

79. A female green rice leafhopper, *Nephotettix cincticeps*.

80. Chlorotic specks and streaks on leaves caused by rice dwarf virus.

81. Close-up of symptoms on leaves caused by rice dwarf virus.

82. Galls on the undersides of leaves caused by rice gall dwarf virus.

83. Galls caused by rice gall dwarf virus, which are at first light green and somewhat translucent, then turn white.

84. Yellowing of the main vein caused by rice giallume virus (cv. Balilla). (Courtesy R. Osler)

85. Browning and necrosis of the leaf sheath caused by rice giallume virus. (Courtesy R. Osler)

86. Yellowing of the leaf blade caused by rice giallume virus. (Courtesy R. Osler)

87. Patches of yellow in a northern Italian rice field caused by rice giallume virus. (Courtesy R. Osler).

88. Adult of *Rhopalosiphum padi*. (Courtesy R. Osler)

89. Nymphs of *Rhopalosiphum padi*. (Courtesy R. Osler)

90. Adult of *Sitobion avenae*. (Courtesy R. Osler)

91. Adult of *Metopolophium dirho dum*. (Courtesy R. Osler)

92. Nymphs of *Metopolophium dirhodum*. (Courtesy R. Osler)

93. Stunting caused by rice grassy stunt virus.

94. A female brown planthopper, *Nilaparvata lugens*.

95. Chlorosis of rice leaves caused by rice hoja blanca.

96. Foliar chlorosis caused by rice hoja blanca. (Courtesy J. M. Vru)

97. Rice plant infected with rice hoja blanca. (Courtesy J. M. Vru)

98. Symptoms of rice necrosis mosaic virus on three leaves (left) compared with a healthy leaf (right). (Courtesy T. Inouye)

99. Serrated leaf margins caused by rice ragged stunt virus.

100. Chlorotic striping on leaves caused by rice stripe virus.

101. Close-up of rice stripe on leaves.

102. A healthy rice plant (left) and a plant infected with rice transitory yellowing virus (right). (Courtesy E. Shikata)

103. A leaf with symptoms of rice transitory yellowing virus (left) and a healthy leaf (right). (Courtesy E. Shikata)

104. A rice field severely infected with tungro.

105. Yellow-orange leaf discoloration caused by infection of rice tungro bacilliform virus and rice tungro spherical virus.

106. Symptoms of yellowing and stunting caused by rice yellow mottle virus on seedlings of various lines (Ahero Rice Research Station, Kisumu, Kenya). Note the apparently resistant line in the foreground. (Courtesy S. N. Fomba)

107. Symptoms of rice yellow mottle virus on a mature plant grown in a tidal swamp of the Great Scarcies River in Sierra Leone. (Courtesy S. N. Fomba)

108. Spikelet malformation and discoloration caused by rice yellow mottle virus. (Courtesy S. N. Fomba)

109. Screening trial for resistance to rice yellow mottle virus in upland rice in Rokupr, Sierra Leone. (Courtesy S. N. Fomba)

110. Yellowing and stunting caused by rice yellow dwarf.

111. Close-up of white tip, caused by *Aphelenchoides besseyi*.

112. White tip in the field, caused by *Aphelenchoides besseyi*. (Courtesy M. D. Miller and R. K. Webster)

113. Close-up of leaves showing the early stages of ufra, caused by *Ditylenchus angustus*.

114. Chlorosis of upper leaves typical of the early stages of ufra.

115. At later stages of ufra, parts of the upper internodes of the stem turn dark brown. (Courtesy J.-C. Prot)

116. Panicle distortion typical of ufra, caused by *Ditylenchus angustus*.

117. Close-up of panicle distortion typical of ufra infestation at a later stage.

118. Damage caused by a severe infestation of *Ditylenchus angustus*.

119. Comparison of plants inoculated with *Hirschmanniella oryzae* (left) and uninoculated plants (right).

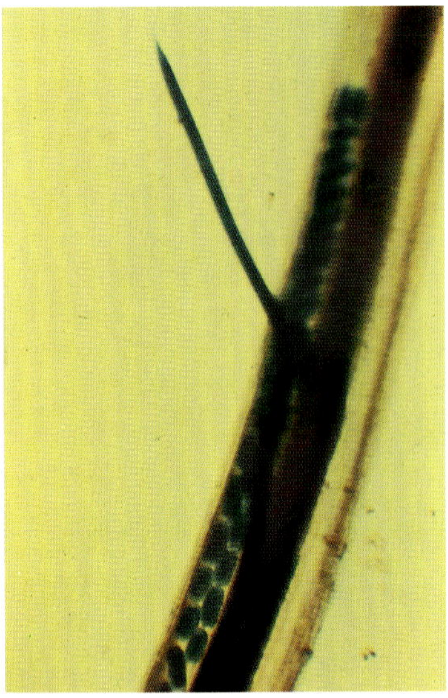

120. Female and eggs of *Hirschmanniella spinicaudata* in a rice root.

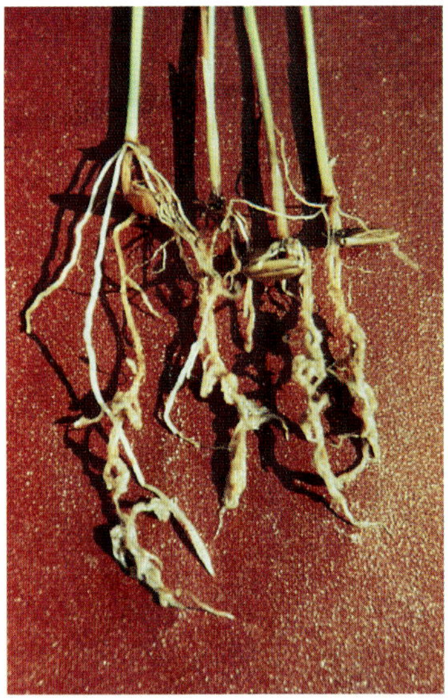

121. Root-knot galls on rice, caused by *Meloidogyne graminicola*.

122. Early akiochi has killed plants, and the remaining plants are declining.

123. Akiochi (H_2S toxicity). Early symptoms consist of decline in plant growth and a black iron sulfide coating on roots as they begin to deteriorate.

124. Straighthead in the field. Note the upright panicles.

125. Close-up of straighthead showing a healthy and a characteristically affected panicle. Note crescent shape of lemmas and paleae, termed parrot beak (left).

126. Zinc deficiency on young rice. Note droopy discolored leaves.

127. Zinc deficiency in water-seeded rice. Note blotches and streaks on droopy, floating leaves on water surface.

128. Zinc deficiency in dry-seeded rice showing typical bronzing of leaves.

129. Nitrogen deficiency. Center plot shows typical light green to yellow symptoms on older, erect leaves. Plot on left is the control with adequate N.

130. Phosphorus deficiency. Plot in foreground shows severe P deficiency with stunted, dead plants. Plot in background is the control with adequate P.

131. Potassium deficiency. Note chlorosis between veins and brown spots near tips of older leaves.

132. Potassium deficiency. Close-up of chlorosis and brown spots between veins and near tips of older leaves.

133. Molinate herbicide damage, showing chlorosis of older leaves and deformed, partially sterile panicles incompletely emerged.

134. Propanil toxicity on rice predisposed to injury by carbofuran treatment. Plot in foreground shows dramatic interaction between a carbamate insecticide and the herbicide propanil. Similar interactions may occur between organic phosphates and propanil. (Courtesy D. M. Brandon and G. B. Trahan)

135. Sulfonylurea herbicide damage. Note chlorotic-necrotic older leaves.

Symptoms

Kernel smut is characterized by the replacement of the kernel endosperm with a black, sooty mass of spores, often termed chlamydospores. Ordinarily, affected plants have only a few smutted grains, which appear at random positions in the panicle (Plate 68). The disease is most noticeable early in the morning, when dew causes completely smutted grains to swell and erupt, releasing the spores. Infected grains may be completely or partially smutted. Partially smutted grains will germinate, but they usually produce stunted seedlings.

Causal Organism

The nomenclature of the kernel smut fungus has been a subject of controversy that has focused on whether the fungus belongs in the genus *Tilletia* or *Neovossia*. We agree with the placement of the fungus in the genus *Neovossia* by Padwick and Khan, but much of the current literature refers to the fungus as *Tilletia barclayana* (Bref.) Sacc. & Syd. in Sacc. Chlamydospores of *Neovossia horrida* (Takah.) Padwick & A. Khan are pulverulent, black in mass, globose to irregularly rounded, sometimes with a short hyaline appendage, 18–23 μm in diameter, and thickly covered with conspicuous spines. The spines are hyaline to slightly colored, pointed at the apex, irregularly polygonal at the base, somewhat curved, and 2.5–4.0 μm long.

The chlamydospores germinate to form a promycelium, which is topped by a single whorl of 20–60 filiform primary sporidia. The primary sporidia do not fuse. Two types of secondary sporidia are formed, an obclavate-allantoid type (15–22 × 5 μm) and a filiform type (35–41 × 2 μm).

Disease Cycle and Epidemiology

Chlamydospores are released from completely smutted grains before or during harvest and contaminate healthy kernels, crop debris, and soil. After planting, chlamydospores float to the surface of the paddy water and germinate to form primary sporidia. Primary sporidia are released from the promycelium and form secondary sporidia in great numbers, either directly or from mycelium produced by the primary sporidia. Secondary sporidia are forcibly discharged into the air and disseminated to rice panicles by air currents. The secondary sporidia can infect the developing ovaries if the florets are open.

Control

Kernel smut is a difficult disease to control. Cultivars differ in susceptibility to the disease, but in general, short-grain cultivars are the most resistant, long-grain cultivars the most susceptible, and medium-grain cultivars are intermediate in resistance. The degree of susceptibility appears to be partly correlated with the duration of anthesis and the size of the floret opening, which is greatest in long-grain cultivars and least in short-grain cultivars. In general, the incidence of kernel smut on semidwarf cultivars is much lower than on tall cultivars. The dense canopy associated with the semidwarfs may protect the florets from sporidia liberated at the water surface.

Heavy applications of nitrogenous fertilizers and excessive flood depths during flowering increase the incidence of disease in fields with a history of infection. No foliar applied fungicides have been found to be effective for controlling kernel smut.

Selected References

Chowdhury, S. 1951. Studies on the bunt of rice (*Oryza sativa* L.). Indian Phytopathol. 4:25-37.

Ou, S. H. 1985. Rice Diseases. 2nd ed. Commonwealth Mycological Institute, Kew, England.

Templeton, G. E. 1963. Kernel smut of rice as affected by nitrogen. Arkansas Farm Res. 12(5):12.

Whitney, N. G. 1977. Effect of fungicide application on kernel smut of rice. Plant Dis. Rep. 61:379-381.

Whitney, N. G. 1989. Taxonomy of the fungus causing kernel smut of rice. Mycologia 81:468-471.

Whitney, N. G., and Frederiksen, R. A. 1975. Kernel smut of rice. Tex. Agric. Exp. Stn. Misc. Publ. 1231.

(Prepared by N. G. Whitney)

Udbatta

Udbatta, also known as black choke, incense rod, and false ergot, has been reported from India, China, Vietnam, Hong Kong, New Caledonia, and West Africa. The disease can cause significant yield losses in areas where it is endemic, but occurrence of the disease is generally sporadic and of minor importance.

Symptoms

Symptoms are usually not apparent until heading, when infected plants produce erect spikes covered by white mycelium instead of normal inflorescences. The spikes eventually become dark and mummified as they are transformed into pseudosclerotia. No grain develops on infected panicles. The flag leaf and sheath of infected tillers are sometimes slightly distorted, and the flag leaf and upper leaves may appear silvery. Infected plants may also be stunted.

Causal Organism

Udbatta is caused by *Balansia oryzae-sativae* Hashioka (anamorph *Ephelis oryzae* Syd.). Hemispherical, black stroma on short stalks develop from the pseudosclerotia (Plate 69). The periphery of the stroma is embedded with ovate to pyriform perithecia, 125–200 × 85–100 μm. Asci are cylindrical, 92–120 × 6 μm, and contain eight ascospores, which are filiform, hyaline, nonseptate, straight to somewhat curved, and 12–27 × 1 μm. Black, cupulate to cushion-shaped, somewhat gelatinous conidiomata also form on the pseudosclerotia. Conidiophores are hyaline and branched and terminate in phialides that produce hyaline, filiform to acicular conidia, 18–22 × 1.5 μm. *Ephelis pallida* has also been described from rice in West Africa, but it is considered to be synonymous with *E. oryzae*.

Disease Cycle and Epidemiology

The disease cycle of udbatta is not fully characterized. *B. oryzae-sativae* is reportedly systemic in infected plants and invades plants at the seedling stage. Infection of panicles by conidia at the time of flowering has also been reported. There is evidence that seedling infections can result from both seedborne and soilborne inoculum. The fungus reportedly attacks several grass weeds, but it is not known if these hosts serve as a source of inoculum for rice. In India, the disease is most prevalent in rice grown at higher elevations (1,500–3,500 m).

Control

The use of clean seed or seed treated with hot water or fungicide has been reported to control the disease. Fungicide soil drenches applied in the nursery bed have also reduced disease. Seed treatment combined with soil treatment reportedly results in better control than either treatment alone. Cultivars vary in susceptibility to the fungus, and resistant cultivars have been identified in India.

Selected References

Booth, C. 1979. *Balansia oryzae-sativae*. Descriptions of Plant Pathogenic Fungi and Bacteria, No. 640. Commonwealth Mycological Institute, Kew, England.

Hashioka, Y. 1971. Rice diseases in the world—VIII. Diseases due to Hypocreales, Ascomycetes (fungal diseases—5). Riso 20:235-258.

Mohanty, N. N. 1975. Efficacy of different physio-chemical methods of seed treatment for control of udbatta disease of rice. Indian Phytopathol. 28:521-523.

Mohanty, N. N. 1977. Mode of infection of udbatta disease of rice.

Riso 26:243-247.
Ou, S. H. 1985. Rice Diseases. 2nd ed. Commonwealth Mycological Institute, Kew, England.
Padhi, N. N., and Mohanty, N. N. 1984. Efficacy of seed and soil treatments with different fungicides on control of udbatta disease of rice. Indian Phytopathol. 37:686-688.

(Prepared by F. N. Lee and P. S. Gunnell)

Glume Blight

Glume blight has been reported from Japan, the United States, Argentina, Brazil, Asia, and Africa. In general, the disease is of minor importance, but under favorable conditions it has caused yield losses of 25–45%.

Symptoms

Symptoms are variable and can occur on the glumes at any time from heading to maturity. The disease may appear as brown, pinhead-sized lesions or as lens-shaped lesions with bleached centers and a brown margins. Under moist conditions, numerous pcynidia develop in the bleached areas. When panicle emergence coincides with a rainy period, the disease appears as reddish brown blotches on the glumes (Plate 70). Infection results in grain sterility and a reduction in grain weight (Plate 71).

Causal Organism

Glume blight is caused by *Phoma sorghina* (Sacc.) Boerema, Dorenbosch & Van Kesteren. This fungus has been reported from rice under several synonyms, including *Phoma glumarum* Ellis & Tracy, *Phoma insidiosa* Tassi, *Phyllosticta glumarum* (Ellis & Tracy) Miyake, *Phyllosticta glumicola* (Speg.) Hara, *Phyllosticta oryzina* Padwick, and *Phyllosticta sorghina* Sacc.

Pycnidia of the fungus are dark brown to black, irregularly globose with a short neck, and 80–200 μm in diameter. Conidia are hyaline, biguttulate or without guttules, ovoid to ellipsoid, and mostly 4–7 × 2–3.5 μm. Single chlamydospores and dictyochlamydospores are formed.

Disease Cycle and Epidemiology

P. sorghina is ubiquitous in tropical and subtropical areas as a saprophyte and a weak pathogen of many gramineous hosts, including sorghum, sugar cane, wheat, maize, millet, and many weed species. The fungus is seedborne and causes the most damage when heading occurs during periods of rain.

Control

The use of clean seed and cultivars that show resistance to the fungus can help control glume blight.

Selected References

Boerema, G. H., Dorenbosch, M. M. J., and Van Kesteren, H. A. 1973. Remarks on species of *Phoma* referred to Pyronellaea—IV. Persoonia 7:131-139.
Ou, S. H. 1985. Rice Diseases. 2nd ed. Commonwealth Mycological Institute, Kew, England.
Prabhu, A. S., and Bedendo, I. P. 1988. Glume blight of rice in Brazil: Etiology, varietal reaction and loss estimates. Trop. Pest Manage. 34:85-88.

(Prepared by F. N. Lee)

Scab

Scab is one of the first recorded diseases of rice and has been reported from most rice-growing regions of the world. The disease is not usually serious but can be severe under favorable conditions.

Symptoms

The fungus initially causes bleached lesions or a bleached discoloration on the glumes. Affected areas become yellow to salmon to carmine with the production of sporodochia and conidial masses of the fungus (Plate 72). Blue-black perithecia may also be present on infected glumes. Infected grains are light, shrunken, and brittle and may be sterile. The fungus has also been reported to infect the nodes of the rice tiller, causing a black rot and resulting in breakage of the stems.

Causal Organism

Scab is caused by *Gibberella zeae* (Schwein.) Petch (anamorph *Fusarium graminearum* Schwabe). Perithecia are superficial, blue-black, ovoid, and 140–250 μm in diameter. Asci are clavate, 60–85 × 8–11 μm, and contain usually eight distichous or obliquely monostichous ascospores. Ascospores are hyaline to light brown, fusoid, somewhat curved, three-septate at maturity, and 19–24 × 3–4 μm. Macroconidia are straight to moderately sickle-shaped, thick-walled, with a distinct foot-shaped basal cell and a short, tapered apical cell, three- to seven-septate, 35–62 × 2.5–3 μm. Globose chlamydospores may be present. Microconidia are absent. On potato-dextrose agar, the fungus produces a distinct carmine pigment and floccose white to tan to yellow mycelium.

Disease Cycle and Epidemiology

The fungus is cosmopolitan and can be found in the soil and on crop residues. Rice is most susceptible to infection at flowering and less susceptible from the milky stage of grain development onward. *G. zeae* also causes scab on wheat.

Control

No control measures are warranted.

Selected References

Nelson, P. E., Toussoun, T. A., and Marasas, W. F. O. 1983. *Fusarium* Species: An Illustrated Manual for Identification. Pennsylvania State University Press, University Park.
Ou, S. H. 1985. Rice Diseases. 2nd ed. Commonwealth Mycological Institute, Kew England.

(Prepared by F. N. Lee)

Black Kernel

At least 14 species of *Curvularia* have been reported from rice. In general, these species are associated with the grain and cause a blackish discoloration, but some may cause leaf spots or seedling blights under certain circumstances.

The species of *Curvularia* that have been reported from *O. sativa* are *C. affinis* Boedijn, *C. clavata* Jain, *C. eragrostidis* (Henn.) J. A. Meyer, *C. fallax* Boedijn, *C. geniculata* (Tracy & Earle) Boedijn, *C. inaequalis* (Shear) Boedijn, *C. intermedia* Boedijn, *C. lunata* (Wakk.) Boedijn, *C. oryzae* Bugnicourt, *C. oryzae-sativae* Sivan., *C. pallescens* Boedijn, *C. senegalensis* (Speg.) Subramanian, *C. tuberculata* Jain, *C. uncinata* Bugnicourt, and *C. verruculosa* Tandon & Bilgrami. The teleomorphs of several of the species have been identified and belong in the genus *Cochliobolus*.

Curvularia species are characterized by conidia that are multicellular, ellipsoid, broadly fusiform, or obovoid, somewhat to strongly curved, with one or more of the center cells larger and darker than the others.

The *Curvularia* species mentioned have been isolated from numerous grass hosts and are regarded as weak pathogens and common saprophytes. Infection of rice kernels in nature may take place at flowering.

Selected References

Ou, S. H. 1985. Rice Diseases. 2nd ed. Commonwealth Mycological Institute, Kew, England.

Sivanesan, A. 1987. Graminicolous species of *Bipolaris, Curvularia, Drechslera, Exserohilum* and their teleomorphs. Commonw. Mycol. Inst. Mycol. Pap. 158.

(Prepared by F. N. Lee)

Minute Leaf and Grain Spot

Nigrospora species are common saprophytes of rice worldwide, occurring on senescent plant parts. On plants weakened by disease, insect pests, or poor nutrition, species of *Nigrospora* may infect the leaves, leaf sheaths, and culms and have been reported to cause an ear blight and blackening of the rice kernels.

Small, powdery, black pustules up to 5 mm in diameter occur on infected plant parts. The fungi enter the host through the stomates and generally grow in the outer parenchyma cell layers.

Species of *Nigrospora* produce smooth, black, opaque, globose to subglobose conidia that are formed singly on short, inflated condiophores. Four species have been reported from rice: *N. oryzae* (Berk. & Broome) Petch, with conidia 10–16 μm in diameter; *N. sphaerica* (Sacc.) E. Mason, with conidia 14–20 μm in diameter; *N. panici* Zimmerm., with conidia 25–30 × 22–25 μm; and *N. padwickii* Prasad, Agnihotri, & Agarwal, with large conidia 33.5–42 × 32–40 μm.

The conidial dimensions of *N. oryzae* and *N. sphaerica* overlap, and there is some question as to whether they are distinct species. Conidia of *N. oryzae* reportedly average 14 μm in diameter, whereas conidia of *N. sphaerica* average 16 μm in diameter. *Khuskia oryzae* H. J. Hudson has been identified as the teleomorph of *N. oryzae*.

Control

No control measures are warranted.

Selected References

Hudson, H. J. 1963. The perfect state of *Nigrospora oryzae*. Trans. Br. Mycol. Soc. 46:355-360.

Ou, S. H. 1985. Rice Diseases. 2nd ed. Commonwealth Mycological Institute, Kew England.

(Prepared by F. N. Lee)

Red Blotch of Grains

In many rice-growing regions, *Epicoccum* species have been reported to cause a pink to rose to red discoloration of rice kernels, known as red blotch. The disease occurs when rice plants lodge before harvest, and the panicles come in contact with the soil. Both immature and mature rice kernels can be infected. Infected seeds fail to germinate or they produce weakened seedlings.

Epicoccum species are common saprophytes and are weak pathogens. *E. oryzae* Ito & Iwad. and *E. neglectum* Desmaz. have both been reported to cause red blotch. Both are synonymous with *E. purpurascens* Ehrenb., which is a later synonym of *E. nigrum* Link.

E. nigrum produces subglobose to pyriform, dark golden brown dictyospores that are heavily warted and measure 15–25 μm in diameter. The spores are produced singly at the apex of short, hyaline conidiophores. On the host, the fungus occurs as small, powdery, brown to blackish sporodochia.

The disease can be avoided by measures that prevent lodging.

Selected References

Booth, C. 1980. *Epicoccum purpurascens*. Descriptions of Plant Pathogenic Fungi and Bacteria, No. 680. Commonwealth Mycological Institute, Kew, England.

Cannon, P. F. 1986. International Commission on the Taxonomy of Fungi (ICTF): Name changes in fungi of microbiological, industrial and medical importance. Part 1. Microbiol. Sci. 6:168-171.

Ou, S. H. 1985. Rice Diseases. 2nd ed. Commonwealth Mycological Institute, Kew England.

(Prepared by F. N. Lee)

Grain Discoloration

Grain discoloration occurs in most rice production regions of the world. Names or terms used to describe discolored rice grain include yellow rice, spotted rice, brown spot, stackburn, red blotch, black kernel, pecky rice, dirty rice, bunt, scab, and black rot.

Symptoms

Discolored grain is often chalky, brittle, and shrunken and has reduced viability (Plate 73). In addition, certain microorganisms associated with discolored grain are known to produce mycotoxins. These factors detract from the quality and appearance of the grain and make it less marketable.

Causal Organisms

Infection and discoloration of the grain may take place before or after harvest. Many of the organisms that can infect grain in the field and cause discoloration are pathogens of other plant parts as well and have been discussed in previous sections. For example, dark discoloration of grain is produced by *Cochliobolus miyabeanus* (Ito & Kuribayashi in Ito) Drechs. ex Dastur, *Gibberella fujikuroi* (Sawada) Ito in Ito & K. Kimura, *G. zeae* (Schwein.) Petch, *Gerlachia oryzae* (Hashioka & Kokogi) W. Gams, *Sarocladium oryzae* (Sawada) W. Gams & D. Hawksworth, *Neovossia horrida* (Takah.) Padwick & A. Khan, and *Alternaria padwickii* (Ganguly) Ellis. A dirty yellow color can result from infection by *Curvularia* spp., *Nigrospora* spp., *Phoma sorghina* (Sacc.) Boerema, Dorenbosch, & Van Kesteren, and *Pseudomonas glumae* Kurita & Tabei. Black spots are produced by *P. fuscovaginae* Tanii, Miyajima, & Akita and *Erwinia herbicola* (Löhnis) Dye. Other microorganisms that discolor grain include *Monascus purpureus* Went, which stains the kernel red, *Wolkia decolorans* (Van der Wolk) Ramsb., which causes yellow grain, and species of *Phoma* and *Phyllosticta*.

Organisms that invade rice grain during storage are primarily species of *Aspergillus* and *Penicillium*, which impart blue, green, and yellow stains. Species of *Absidia, Mucor, Rhizopus, Chaetomium, Dematium, Monilia, Streptomyces, Syncephalastrum*, and *Verticillium* are also found on stored rice as well.

Disease Cycle and Epidemiology

Rainfall and/or high humidity during grain production favor grain discoloration. Also, lodging of the rice plant, which places the panicle in a wet or humid environment, increases grain discoloration. In storage, grain with a moisture content above 14.5% allows for invasion by microorganisms. Also, wounds caused by feeding insects on rice grains in the field or in storage can provide entry sites for microorganisms. Ingress of

Nematospora coryli Peglion, *Curvularia* spp., *Alternaria* spp., *Fusarium* spp., and *Bipolaris* spp. are associated with feeding by the rice stink bug, *Oebalus pugnax* (Fabricius). The damage caused by the rice stink bug and the associated fungi is known as pecky rice.

Control

Control measures for foliar and sheath pathogens that can also cause grain discoloration have been outlined in previous sections. Proper insect control and cultural practices that prevent lodging and will also help minimize grain discoloration. In addition, it is crucial that grain be harvested at the proper moisture content or dried to prevent invasion of microorganisms in storage.

Selected References

Lee, S. C., Alvenda, M. E., Monman, J. M., and Heinrichs, E. A. 1986. Insects and pathogens associated with rice grain discoloration and their relationship in the Philippines. Korean J. Plant Prot. 25:107-112.

Ou, S. H. 1985. Rice Diseases. 2nd ed. Commonwealth Mycological Institute, Kew, England.

Padwick, O. W. 1950. Manual of Rice Diseases. Commonwealth Mycological Institute, Kew, England.

Tullis, E. C. 1936. Fungi Isolated from Discolored Rice Kernels. Tech. Bull. U.S. Dep. Agric. 540.

(Prepared by F. N. Lee)

Ear Blight

Ear blight refers to a disease complex that involves several fungi. The disease is also known as ear blighting, ear or glume discoloration, panicle blight, and false blast. The disease has been reported from Japan, Taiwan, Thailand, India, Korea, and the United States.

Symptoms

Symptoms of ear blight vary considerably, depending on the fungi involved, environmental conditions, plant age and nutritional status, and cultivar susceptibility. In general, ear blight is characterized by a discoloration and/or blight of the entire panicle or portions of the panicle, including the neck, branches, internodes, nodes, and spikelets (Plate 74). The discoloration may be defined, occurring as brown to black streaks, or it may appear effuse. The grain of affected panicles often develops poorly or aborts.

Causal Organism

Many fungi have been associated with ear blight, including some that cause major foliar diseases. Fungi that have been cited are *Bipolaris oryzae* (Breda de Haan) Shoemaker, *Cercospora oryzae* Miyake, *Curvularia lunata* (Wakk.) Boedijn, *Fusarium nivale* (Fr.) Ces., *Fusarium* spp., *Alternaria oryzae* K. Hara, *Phoma* spp., *Gerlachia oryzae* (Hashioka & Yokogi) W. Gams, *Nakataea sigmoidea* (Cavara) K. Hara, *Cladosporium* spp., *Epicoccum* sp., and *Nigrospora* spp.

Disease Cycle and Epidemiology

Variables affecting the occurrence of the major leaf pathogens directly influence the incidence and severity of ear blight. However, ear blight is more likely to occur in late-planted fields or in fields deficient in nitrogen, potassium, iron, manganese, and magnesium. Heavy rainfall during heading and extreme temperatures during maturation increase its severity.

Control

The disease can best be prevented by following recommended control procedures for the major leaf pathogens involved. Plantings should be timed to mature before cool weather is likely to begin, and fields should have a proper balance of nutrients sufficient for normal plant development. Fungicides are reported to provide some control.

Selected References

Ezuka, A., and Watanabe, Y. 1973. Summarized results of the joint study on rice ear blighting conducted by seven agricultural experiment stations in Tokai-Kinki Region. (In Japanese; English summary.) Proc. Kansai Plant Prot. Soc. 15:37-41.

Kang, C. S., and Kim, C. K. 1972. Studies on the fungi associated with ear blight of rice. (In Korean; English summary.) Korean J. Plant Prot. 11:101-107.

Ohata, K. I. 1978. Studies on panicle blight of rice plants. (In Japanese.) Ann. Phytopathol. Soc. Jpn. 44:235-237.

Ou, S. H. 1985. Rice Diseases. 2nd ed. Commonwealth Mycological Institute, Kew, England.

(Prepared by F. N. Lee)

Obscure Fungal Diseases of the Foliage and Glumes

Many obscure fungal diseases of the leaves and glumes have been reported. Often infection was noted when the plants were weak, dying, or under adverse environmental conditions, and it is likely that some of the fungi involved are extremely weak pathogens or saprophytes. Most reports concerning these fungi come from Japan, China, Taiwan, the Philippines, India, Italy, and Burma. Very little information concerning the diseases they cause is available or has been corroborated, and the validity of some of the fungal species themselves is in question. Therefore, simply a list of the fungi that have been reported is provided here.

Alternaria oryzae Hara
Bipolaris
  australiensis (M. B. Ellis) Tsuda & Ueyama
  hawaiiensis (M. B. Ellis) Uchida & Aragaki
  spicifera (Bainier) Subramanian
Cladosporium
  herbarum (Pers.:Fr.) Link
  miyakei Sacc. & Trott.
  oryzae Sacc. & Syd.
Coniothyrium
  anomalum Miyake
  brevisporum Miyake
  japonicum Miyake
Diplodia oryzae Miyake
Exserohilum rostratum (Drechs.) Leonard & Suggs
Helococera oryzae Linder & Tullis (*Helococera* is a nomen confusum)
Melanomma glumarum Miyake
Metasphaeria
  culmicola (Fr.) Auersw.
  culmifraga (Fr.) Ces. & De Not.
  inecola Hara
  michotti (West.) Sacc.
  oryzae (Cattaneo) Sacc.
  oryzicola Hara
Mycosphaerella
  graminicola (Fuckel) Oudem.
  hondai (Miyake) Tomilin

malinverniana (Cattaneo) Jacz.
oryzae (Garov. & Cattaneo) Jacz.
tulasnei (Jancz.) Lindau.
usteriana (Speg.) Hara
Oospora (Oospora is an illegitimate name for *Oidium)*
oryzae Ferraris
oryzetorum Sacc.
Pestalotia
kawakami Sawada
oryzae Hara
Phaeoseptoria
japonica Hara
oryzae Miyake
tassiana (De Not.) Johans.
Sphaeropsis
japonicum Miyake
oryzae (Cattaneo) Sacc.
vaginarum (Cattaneo) Sacc.
Trematosphaerella
cattanei (Thuem.) Padwick
oryzae (Miyake) Padwick

(Prepared by F. N. Lee and P. S. Gunnell)

Diseases Caused by Viruses and Mycoplasmalike Organisms

Since the mid-1950s, virus diseases and yellow dwarf, which is caused by a mycoplasmalike organism (MLO), have become increasingly important in many rice-growing areas. This is especially true in the tropics, where virus diseases have become a major constraint to stable rice production. The increase in incidence and severity of virus and MLO diseases can be attributed to the development and widespread adoption of new, "improved" rice cultivars that are typically short-statured and mature early. The use of early-maturing cultivars has intensified rice production in many areas by allowing two crops per year, thus favoring the buildup of viruses and MLOs and their insect vectors. The use of short-statured cultivars allows for increased use of nitrogen fertilization to obtain maximum yields, and the erect growth habit of the improved cultivars allows for reduced space between plants, making the rice stands a more favorable environment for insect vectors. In addition, many of the improved cultivars are more susceptible to the viral and MLO agents or their vectors than traditional or land race cultivars. Furthermore, individual improved cultivars are genetically uniform, and many of them have been derived from similar genetic backgrounds. Such genetic uniformity, coupled with widespread deployment of these cultivars, has exerted intense selection pressure on the virus, MLO, and vector populations for virulent strains and has favored disease spread.

Fourteen of the 16 major viruses and both MLOs that cause rice diseases are insect vectored (Table 6). Rice necrosis mosaic virus and rice stripe necrosis virus are the exceptions, being vectored most likely by a soilborne fungus. Leafhoppers and planthoppers vector the majority of the viruses and MLOs. Most of the viruses are circulative and propagative in their insect vectors, and several are passed transovarially.

Control of rice diseases caused by viruses and MLOs is often difficult. Disease incidence is generally correlated with vector populations, but reducing vector density is problematic. In warm areas, insects that harbor and transmit the disease agents can be present year-round. In areas where vectors cannot overwinter, migrations of vectors such as the brown planthopper can introduce disease agents from up to 1,000 km away. Applications of insecticides aimed at resident vector populations or influxes of migratory vectors are difficult to time and are often inadequate to reduce vector populations to levels that do not pose a threat. Also, insecticides used for vector control reduce populations of beneficial insects that parasitize or prey on vectors, and some vector populations have developed resistance to the insecticides used. Certain cultural practices may help reduce vector populations—such as decreasing plant density, proper timing of planting, and synchronous planting of rice crops—but may not be effective or feasible in all situations. Host resistance to the vector and/or the virus or MLO is the most practical and effective means of disease control, but it has not always proved durable.

Virus and MLO diseases are most destructive when the rice plants are infected at the seedling stage. Generally, plants that are infected during the early growth stages develop little grain, and total yield loss is common with severe infections. Plants infected at later growth stages often develop mild symptoms or no symptoms.

Diagnosis of rice diseases caused by viruses and MLOs based on symptomatology is difficult, because many of the agents produce similar symptoms and because mixed infections can occur. Also, symptoms caused by viruses and MLOs can mimic physiological disorders and diseases caused by other pests. Proper diagnosis of virus diseases is largely dependent on serological techniques that detect the presence of the viral proteins.

(Prepared by H. Hibino)

Virus Diseases

Black-Streaked Dwarf

Black-streaked dwarf occurs in China, Japan, and Korea. Occurrence of the disease is generally sporadic and limited to small patches in a field, but major outbreaks of the disease were recorded in Japan in 1941 and in Japan and China during the 1960s.

Symptoms

Symptoms appear 14–24 days after infection. Early symptoms include stunting, dark green discoloration of the leaves, and occasional vein swellings on the sheaths (Plate 75). Infected plants may also show twisting of the leaf tips and splitting of the leaf margins. At later growth stages, prominent dark brown streaks consisting of swollen veins or galls form along the undersides of leaves and on the sheaths and culms (Plate

76). The swellings and galls are initially white and result from the proliferation of phloem tissue. Infected plants produce no panicles or poorly formed panicles that are only partially exserted. Dark brown blotches may form on the kernels. The virus causes similar symptoms on other commercial grass hosts but is most damaging to rice and maize. Symptoms of rice black-streaked dwarf are somewhat similar to those of rice ragged stunt and rice gall dwarf viruses.

Causal Agent

Rice black-streaked dwarf virus (RBSDV) is a member of the fijivirus group (Reoviridae). The virus particles are isometric and consist of a double shell, composed of six proteins, that has spikes or projections at each vertex (Fig. 6). The genome is composed of 10 double-stranded RNA segments. In the plant host, the virus particles are restricted to the phloem and gall cells. Complete particles occur in the

TABLE 6. Diseases of Rice Caused by Viruses and Mycoplasmalike Organisms

	Vectors[a]	Vector Relationship	Geographical Distribution
Phytoreoviruses			
Dwarf	*Nephotettix cincticeps* (Uhler)* *N. nigropictus* (Stål) *N. virescens* (Distant) *Recilia dorsalis* (Motschulsky)	Circulative, propagative, transovarial	China, Japan, Korea, Nepal
Gall dwarf	*N. nigropictus** *N. cincticeps* *N. malayanus* Ishihara & Kawase *N. virescens* *R. dorsalis**	Circulative, propagative, transovarial	China, Malaysia, Thailand
Fijiviruses			
Black-streaked dwarf	*Laodelphax striatellus* (Fallen)* *Unkanodes sapporona* (Matsumura) *U. albifascia* Matsumura	Circulative, propagative	China, Japan
Ragged stunt	*Nilaparvata lugens* (Stål)* *N. bakeri* Muir	Circulative, propagative	South and Southeast Asia, China, Japan, Taiwan
Rice stripe viruses			
Stripe	*L. striatellus** *U. sapporona* *U. albifascia* *Terthron albovittatus* (Matsumura)	Circulative, propagative, transovarial	China, Japan, Korea, USSR
Hoja blanca	*Sogatodes orizicola* (Muir)* *S. cubanus* (Crawford)	Circulative, propagative	South and Central America
Grassy stunt	*Nilaparvata lugens** *N. bakeri* *N. muiri* China	Circulative, propagative	South and Southeast Asia, China, Japan, Taiwan
Luteoviruses			
Giallume	*Rhopalosiphum padi* (Linnaeus)* *Sitobion avenae* (Fabricius) *Metopolophium dirhodum* (Walker)	Circulative	Italy
Rhabdoviruses			
Transitory yellowing	*Nephotettix cincticeps** *N. nigropictus** *N. virescens**	Circulative, propagative	China, Japan, Taiwan, Thailand
Sobemoviruses			
Yellow mottle	*Apophylia* spp. *Chaetocnema* spp. *Sesselia pusilla* Gerstaecker *Trichispa sericea* Guerin	Noncirculative	Africa
Unclassified viruses			
Tungro	*Nephotettix virescens** *N. cincticeps** *N. nigropictus* *N. malayanus* *N. parvus* Ishihara *R. dorsalis*	Noncirculative	South and Southeast Asia, China, Japan
Necrosis mosaic	*Polymyxa graminis* Ledingham	...	India, Japan
Stripe necrosis	*P. graminis*	...	Ivory Coast
Bunchy stunt	*Nephotettix virescens** *N. cincticeps**	Circulative	China
Mycoplasmalike organisms			
Orange leaf	*R. dorsalis*	Circulative	South and Southeast Asia
Yellow dwarf	*Nephotettix cincticeps** *N. nigropictus** *N. virescens** *N. parvus* *N. malayanus*	Circulative, propagative	China, Asia

[a] Asterisks denote principal vectors.

cytoplasm and measure 75–85 nm in diameter. Incomplete particles, which lack the outer shell, occur in viroplasms in the cytoplasm and measure 50–55 nm in diameter. RBSDV is closely related serologically to maize rough dwarf, cereal tillering disease, and pangola stunt viruses.

Vectors

RBSDV is transmitted primarily by *Laodelphax striatellus* (Fallén), but also by *Unkanodes sapporona* (Matsumura) and *U. albifascia* Matsumura (Plate 77). RBSDV is circulative and propagative in the vector planthoppers. The planthoppers acquire RSBDV in a 12-hr acquisition access period (minimum 30 min). The latent period of RBSDV in the vectors is 7–35 days (average 15–18 days). The vectors can transmit the virus to rice in a 3–6 hr inoculation access period (minimum 5 min) and remain infective until death. Within a population, 50–90% of *L. striatellus*, 34–45% of *U. sapporona*, and 30–73% of *U. albifascia* are active transmitters of the virus. *L. striatellus* can transmit RBSDV from maize to rice, but the efficiency of transmission is very low.

Disease Cycle and Epidemiology

RBSDV persists primarily in grass hosts, including maize, wheat, barley, oats, and several weeds, notably *Alopecurus aequalis* Sobol. After harvest of rice and maize, adult planthoppers move to grass weeds, wheat, and barley, where they oviposit. The planthoppers overwinter as diapausing nymphs on the grass hosts. First-generation adults from the overwintered population acquire RBSDV from infected grasses and transmit the virus to rice in early summer. An outbreak of RBSDV in Shanghai in 1966 was correlated with the presence of large acreages devoted to wheat cultivation.

Disease incidence and severity is greatest in rice planted at the time the first-generation adults disperse from the overwintering hosts, because the density of the first-generation adults is greater than that of subsequent generations, and the rice plants are most susceptible to RBSDV at the seedling stage.

Control

Certain Indica cultivars of rice are resistant to RBSDV, but no high level of resistance is available in Japonica cultivars or in maize and other cereals. Certain tolerance genes have been incorporated into rice breeding lines, however, from cultivars that develop only mild symptoms when infected with RBSDV. Applications of insecticides that lower the population of the first-generation planthoppers can help reduce disease incidence. Late planting also decreases RBSDV incidence by avoiding first-generation adults from winter grasses and cereals.

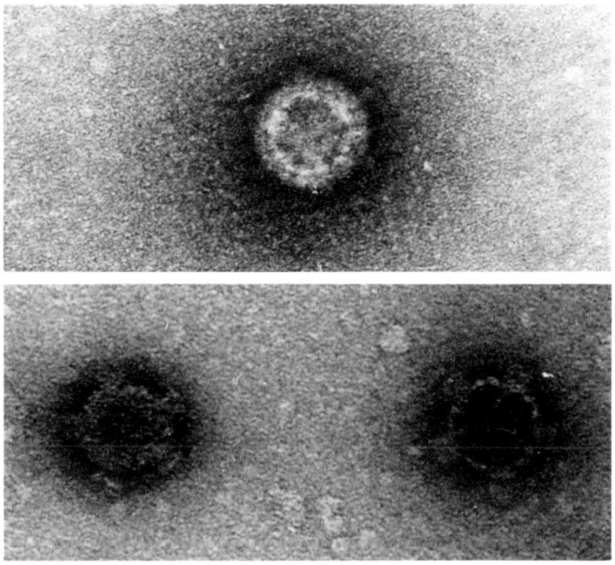

Fig. 6. Top, An intact particle of rice black-streaked dwarf virus; Bottom, two spiked cores of the virus (×240,000). (Courtesy E. Shikata)

Selected References

Ishii, M., and Yoshimura, S. 1973. Epidemiological studies on rice black-streaked dwarf virus in Kanto-Tosan District in Japan. (In Japanese; English summary.) J. Cent. Agric. Exp. Stn. 17:61-121.

Milne, R. C., and Luisoni, E. 1977. Serological relationships among maize rough dwarf-like viruses. Virology 80:12-20.

Reddy, D. V. R., Shikata, E., Boccardo, G., and Black, L. M. 1975. Co-electrophoresis of double-stranded RNA from maize rough dwarf and rice black-streaked dwarf viruses. Virology 67:279-282.

Shikata, E. 1974. Rice black-streak dwarf virus. Descriptions of Plant Viruses, No. 135. Commonwealth Mycological Institute and Association of Applied Biologists, Kew, England.

Shikata, E., and Kitagawa, Y. 1977. Rice black-streaked dwarf virus: Its properties, morphology and intracellular localization. Virology 77:826-842.

(Prepared by H. Hibino)

Bunchy Stunt

Bunchy stunt has been reported in Fujian, Guandong, Hunan, and Jiangxi provinces in China. The etiology of the disease is not well understood.

Symptoms

Infected plants are stunted and tiller profusely. The leaves are short and narrow, similar to those of plants infected with rice dwarf virus (RDV), and young leaves may show mottling. In severely affected cultivars, branches are formed from the upper nodes, and leaves are produced in bunches.

Causal Agent

Bunchy stunt is suspected to be caused by a virus. Viruslike particles, 30–60 nm in diameter, have been found in the phloem of symptomatic plants. Infected plants are not cross-protected from infection by RDV, and antiserum to RDV does not react with plant extracts infected with bunchy stunt.

Vectors

The suspected virus is transmitted by the green leafhoppers, *Nephotettix cincticeps* (Uhler) and *N. virescens* (Distant), and is circulative in the vectors (Plates 78 and 79). The leafhoppers acquire the bunchy stunt agent in a 12-hr acquisition access period (minimum 5 min) and become infective after a latent period of 8–25 days (average 12 days). The percentage of active transmitters in a population of the vector species ranges from 4 to 39% (average 13%) for *N. cincticeps* and from 3 to 11% (average 8%) for *N. virescens*.

Disease Cycle

Rice is the only known host of the bunchy stunt agent. Leafhoppers acquire the agent from ratoons of overwintered rice plants infected with the virus.

Selected References

Xie, L. H., and Lin, H. Y. 1980. Studies on bunchy stunt disease of rice, a new virus disease of rice plants. Kexue Tongbao 25:785-789.

Xie, L. H., Lin, Q. Y., and Zhu, Q. L. 1982. On the bunchy stunt disease of rice. II. The distribution, loss, hosts, and overwintering of rice bunchy stunt. (In Chinese; English summary.) Acta Phytopathol. Sin. 12:16-20.

(Prepared by H. Hibino)

Dwarf

Rice dwarf occurs in Japan, China, Nepal, and Korea. The disease was first recorded in Japan in 1880. Rice dwarf is an important virus in the history of plant virology, as it was the first plant virus demonstrated to multiply in an insect and the first shown to be transmitted and passed transovarially by an insect.

Symptoms

Infected plants show pronounced stunting and produce diminutive tillers or a reduced number of tillers. Root growth is severely retarded. Irregular chlorotic specks are formed on the leaves and sometimes on the leaf sheaths (Plates 80 and 81). The specks may fuse to form broken streaks that parallel the leaf veins. Mature plants remain more or less green at maturity and either do not form panicles or form poorly developed panicles. Symptoms become evident 8–10 days after inoculation at the seedling stage and 27 days after inoculation at the 13th-leaf stage. Plants infected at later stages may show no symptoms.

Causal Agent

Rice dwarf virus (RDV) is a member of the phytoreovirus group (Reoviridae). RDV differs from all other plant viruses in the Reoviridae, because it does not induce hyperplasia of the plant host cells and because it is not restricted to phloem cells, but invades other vascular tissue and adjacent mesophyll cells. Virus particles are isometric, 70 nm in diameter; the capsid consists of a double shell and is composed of seven proteins (Fig. 7). The genome is divided into 12 double-stranded RNA segments. Antiserum to disassociated RDV reacts with one of the proteins of rice gall dwarf virus.

In infected plants, RDV particles occur in viroplasms or in tubular structures in the cytoplasm. In the vector, RDV particles often occur in close association with symbionts in the mycetocyte.

Vectors

RDV is transmitted by *Nephotettix cincticeps* (Uhler), *N. nigropictus* (Stål), *N. virescens* (Distant), and *Recilia dorsalis* (Motschulsky). *N. cincticeps* is the major vector in Japan, Korea, and central China. RDV is circulative and propagative and is passed transovarially. The vectors acquire RDV in a 5-hr acquisition access period (minimum 1 min). The latent period of RDV in the vectors ranges from 4 to 58 days (average 12–35 days) for *N. cincticeps* and from 9 to 42 days (average 10–15 days) for *R. dorsalis*. The vectors can transmit RDV to rice during an inoculation access period of 30 min to 1 hr (minimum 3–10 min) and remain infective for life. The percentage of active transmitters in populations of the vectors ranges from none to 70% for *N. cincticeps* and from none to 40% for the other vector species.

The progeny of infective females are congenitally infected with the virus, and RDV can be passed from a single female of *N. cincticeps* through six generations. RDV is most efficiently passed transovarially by *N. cincticeps*. Nymphs of *N. cincticeps* from an infective female can transmit the virus within 1–19 days after hatching.

Infective leafhoppers have a shorter life span, reduced fecundity, greater nymphal mortality, and shorter oviposition periods.

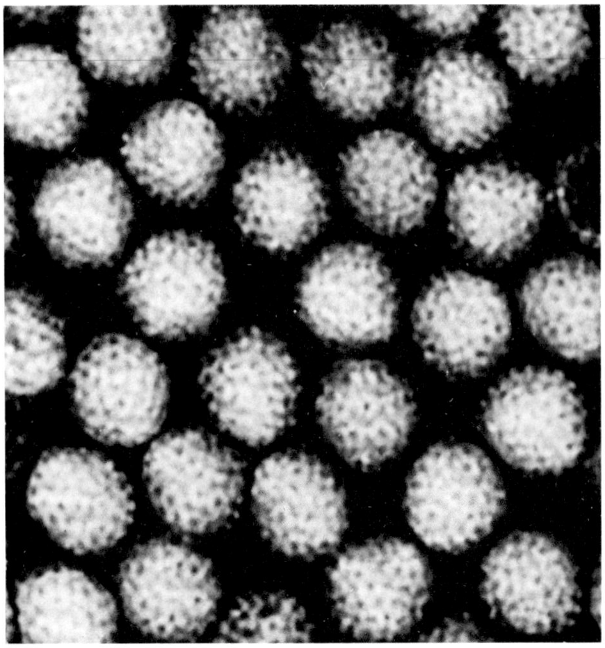

Fig. 7. Rice dwarf virus particles from a purified preparation (×200,000). (Courtesy I. Kimura)

Disease Cycle and Epidemiology

RDV persists primarily in rice and the leafhopper vectors. Although a few grass weeds are susceptible to RDV, they are probably not important as a reservoir of the virus. Infective leafhoppers overwinter as nymphal diapauses on winter cereals and gramineous weeds and move to rice as adults. In areas where rice is planted early, the adults from overwintered populations move directly to rice. In the absence of rice, the adults remain on weed hosts, and the first-generation adults move to rice planted midseason. Epidemics of RDV in Japan from 1955 to 1974 were correlated with the introduction of early-planted rice, which increased the vector density and the proportion of infected vectors. In China, infective second-generation leafhoppers disperse the virus from early- to late-planted rice crops, and RDV incidence is highest in late-planted rice.

Control

Since 1974, the incidence of RDV in Japan has declined with the advent of machine transplanting. The seedlings for mechanical planting are grown in trays in incubators, where they are protected from insect vectors. Also, insecticides are applied to the seedlings prior to planting. However, outbreaks of the virus have occurred because insecticide use has reduced populations of beneficial spiders that are predators of the leafhopper nymphs, and resistance to certain insecticides has developed in the vector population.

Cultural practices that eliminate or reduce overwintering sites for the vector, such as planting fallow rice fields with nonvector hosts or ploughing fallow paddy fields, can help lower RDV incidence. Also, late planting, synchronous planting, or avoiding an overlap of early- and late-planted rice crops should help reduce RDV incidence. In addition, cultivars that are tolerant of RDV or the vectors are available.

Selected References

Iida, T. T., Shinkai, A., and Kimura, I. 1972. Rice dwarf virus. Descriptions of Plant Viruses, No. 102. Commonwealth Mycological Institute and Association of Applied Biologists, Kew, England.

Ishii, M., Yasuo, S., and Yamaguchi, T. 1970. Epidemiological studies on rice dwarf disease in Kanto-Tosan District, Japan. (In Japanese; English summary.) J. Cent. Agric. Exp. Stn. 14:1-115.

Matsuoka, M., Minobe, Y., and Omura, T. 1985. Reaction of antiserum against SDS-dissociated rice dwarf and a polypeptide of rice gall dwarf virus. Phytopathology 75:1125-1127.

Miyai, S., Kiritani, K., and Nakasuji, F. 1986. Models of epidemics of rice dwarf. Pages 459-480 in: Plant Virus Epidemics: Monitoring, Modelling and Predicting Outbreaks. G. D. McClean, R. G. Garrett, and W. G. Ruesink, eds. Academic Press, New York.

(Prepared by H. Hibino)

Gall Dwarf

Gall dwarf was first recognized in Thailand in 1980. Since then the disease has been reported in Malaysia and in Southern China, where it reached epidemic levels in Guandong Province in 1981–1982.

Symptoms

Typical symptoms of the disease are pronounced stunting, dark green leaves, twisting of the leaf tips, and small galls on the underside of leaf blades and on the outer surface of leaf sheaths (Plates 82 and 83). The galls result from hyperplasia of the phloem tissue and measure 0.4–2 × 0.5 mm. The galls are at first light green and somewhat translucent, becoming white, but not dark brown like those caused by rice black-streaked dwarf virus. Infected plants produce few tillers, remain green at maturity, and form a few poorly developed panicles that are not fully exserted.

Causal Agent

Rice gall dwarf virus (RGDV) belongs to the phytoreovirus group (Reoviridae) and is a double-shelled, isometric virus approximately 65 nm in diameter (Fig. 8). The shells are composed of seven proteins, and the genome consists of 12 double-stranded RNA segments. In the host, the virus is restricted to the phloem and gall cells. Antiserum to disassociated, but not intact, rice dwarf virus reacts with one of the proteins of RGDV. Rice dwarf virus is distinct from RGDV, however, in symptomatology and is not restricted to the phloem cells.

Vectors

The most efficient vectors of RGDV are *Nephotettix nigropictus* (Stål) and *Recilia dorsalis* (Motschulsky), but the virus is also transmitted by *N. cincticeps* (Uhler), *N. malayanus* Ishihara & Kawase, and *N. virescens* (Distant). RGDV is circulative and propagative and is passed transovarially in the leafhopper vectors. Leafhoppers acquire the virus in a 12-hr acquisition access period (minimum 3 hr) and can transmit the virus after a latent period of 5–24 days (average 15–18 days). The minimum inoculation access period is 1 hr. The percentage of active transmitters in populations of the vector species ranges from 2 to 95% for *N. nigropictus*, 4 to 17% for *R. dorsalis*, 1 to 43% in *N. cincticeps*, and 0.1 to 14% in *N. virescens*; it is 9% in *N. malayanus*.

Disease Cycle and Epidemiology

RGDV persists in rice, the grass host *Alopecurus aequalis* Sobol., and the vector leafhoppers. *Oryza rufipogon* Griff. and several cereals have been infected experimentally, but natural infection of these hosts has not been documented. In southern China, ratoons from overwintered rice infected with RGDV and *A. aequalis* provide a reservoir of virus inoculum in the spring. The percentage of RGDV-infective individuals in overwintering populations of the vectors is 2–10% for *R. dorsalis*, 0–1% for *N. cincticeps*, and none for *N. nigropictus*.

Selected References

Inoue, H., and Omura, T. 1982. Transmission of rice gall dwarf virus by the green rice leafhopper. Plant Dis. 66:57-59.

Matsuoka, M., Minobe, Y., and Omura, T. 1985. Reaction of antiserum against SDS-dissociated rice dwarf virus and a polypeptide of rice gall dwarf virus. Phytopathology 75:1125-1127.

Omura, T., Hibino, H., Inoue, H., and Tsuchizaki, T. 1985. Particles of rice gall dwarf virus in thin sections of diseased rice plants and insect vector cells. J. Gen. Virol. 66:2581-2587.

Omura, T., and Inoue, H. 1985. Rice gall dwarf virus. Descriptions of Plant Viruses, No. 296. Commonwealth Mycological Institute and Association of Applied Biologists, Kew, England.

Omura, T., Morinaka, T., Inoue, H., and Saito, Y. 1982. Purification and some properties of rice gall dwarf virus, a new phytoreovirus. Phytopathology 72:1246-1249.

(Prepared by H. Hibino)

Giallume

Giallume occurs in northern Italy, where it was a serious problem from 1966 to 1980. Giallume is caused by a strain of barley yellow dwarf virus (BYDV). BYDV is distributed worldwide, but infection of rice by a strain of BYDV has not been reported outside of Italy.

Symptoms

Leaves of infected plants are yellow to orange and may become necrotic (Plate 84). Leaves also are smaller, erect, and may have serrated margins (Plates 85 and 86). Infected plants are stunted, have a reduced number of tillers, and produce blasted florets. Severely infected plants are killed. Symptoms develop approximately 12–20 days after infection. Yield losses from 5 to 100% have been reported in Italy (Plate 87).

Causal Agent

The giallume virus is a member of the luteovirus group and is closely related serologically to the strain of BYDV vectored by *Rhopalosiphum padi* (Linnaeus). Virus particles are isometric, 20–25 (usually 25) nm in diameter, and the genome consists of a single strand of RNA.

Vectors

R. padi is the primary vector of rice giallume, but the virus is also vectored by *Sitobion avenae* (Fabricius) and *Metopolophium dirhodum* (Walker) (Plates 88–92). The virus is circulative but probably does not multiply in the insect vector. *R. padi* acquires the virus after a 2-hr acquisition access period. After a latent period, the aphids can transmit the virus in a 1- to 2-hr inoculation access period. Aphids may remain infective for life.

Disease Cycle and Epidemiology

The rice giallume virus persists in cereals, grass weeds, and the aphid vector. A common perennial grass in and around rice fields in Italy, *Leersia oryzoides* (L.) Sw., serves as a reservoir for the virus and a host for *R. padi*. *R. padi* is attracted to *L. oryzoides* because it emerges from the paddy water before the rice seedlings. Distribution of giallume is often patchy and associated with *L. oryzoides* in the rice fields.

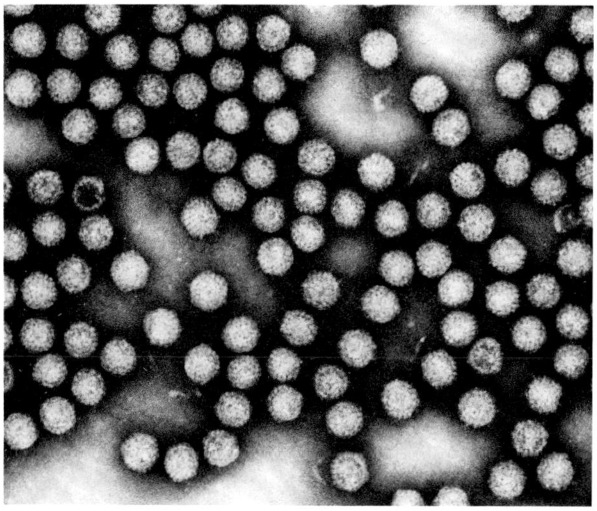

Fig. 8. Rice gall dwarf virus particles from a purified preparation (×75,000). (Courtesy T. Omura)

Control

The use of resistant cultivars coupled with practices aimed at reducing populations of *L. oryzoides* can limit the disease. Insecticides can help reduce the vector density but are not recommended unless highly susceptible cultivars are being grown. Parasites of *R. padi* can help keep the vector population in check.

Selected References

Osler, R. 1984. Rice giallume, a disease related to barley yellow dwarf in Italy. Pages 125-131 in: Barley Yellow Dwarf: A Proceedings of the Workshop. Centro Internacional de Mejoramiento de Maiz y Trigo (CIMMYT), Mexico City.

Osler, R., Amici, A., and Belli, G. 1974. Transmission of rice "giallume" by an aphid *Rhopalosiphum padi*. Riv. Patol. Veg. 10:3-15.

Osler, R., Amici, A., and Longoni, C. E. 1980. *Leersia oryzoides*, a natural host and winter reservoir of the rice "giallume" strain of barley yellow dwarf virus. Phytopathol. Z. 97:242-251.

(Prepared by H. Hibino)

Grassy Stunt

Grassy stunt is widely distributed in the rice-growing areas of South and Southeast Asia. It also occurs in southern China, southern Japan, and Taiwan. Before 1970, grassy stunt occurrence was sporadic, but from 1972 to 1978 severe outbreaks of the disease were recorded in India, Indonesia, Thailand, and the Philippines in association with high infestations of the brown planthopper vector. During the 1980s, disease incidence was low in these countries.

Symptoms

Infected plants show pronounced stunting, proliferation of tillers, and short, narrow, pale green to pale yellow leaves (Plate 93). Newly expanded leaves may be mottled or striped. Leaves also may have numerous small, irregular, dark brown or rust-colored spots. Infected plants produce no panicles or a few panicles with unfilled kernels. A strain of the virus in the Philippines, designated as rice grassy stunt virus-2, mimics tungro symptoms by causing yellow to orange discoloration of the leaves. A severe strain of the virus in Taiwan that causes yellowing of the leaves and wilting is known as rice wilted stunt virus. Symptoms become evident 10–20 days after infection.

Causal Agent

Rice grassy stunt virus (RGSV) is similar to viruses in the tenuivirus (rice stripe virus) group. RGSV particles are fine, flexuous filaments 6–8 nm in diameter. Purified particles are often circular, with a modal contour length of 950–1,350 nm. RGSV particles have one capsid protein (molecular weight 31,000), and the genome consists of four single strands of RNA, which are probably of both positive and negative senses. RGSV is distantly related serologically to rice stripe virus.

Vectors

The major vector of RGSV is the brown planthopper, *Nilaparvata lugens* (Stål) (Plate 94). The virus is also transmitted by *N. bakeri* Muir and *N. muiri* China. Unlike other viruses in the tenuivirus group, RGSV is not passed transovarially in the vectors. Planthoppers acquire RGSV in a 6-hr acquisition access period (minimum 30 min). After a latent period of 5–25 days (average 10–11 days), the planthoppers can transmit the virus in an inoculation access period of several minutes to 25 hr (minimum 5–15 min). Viruliferous insects remain infective for life.

Disease Cycle and Epidemiology

RGSV persists in rice and the vector planthoppers. In the tropics, grassy stunt is generally endemic where rice is grown year-round. Migratory flights of the planthoppers are responsible for disseminating the virus to summer-grown rice in temperate regions.

Control

Resistance to RGSV from *Oryza nivara* Sharma & Shastry has been incorporated into cultivars of *O. sativa* L., and such cultivars have been widely deployed. However, in several countries, strains of RGSV have developed that overcome the resistance. Rice cultivars resistant to the brown planthopper exhibit a low incidence of RGSV, but in some areas, this type of resistance has also broken down as a result of the development of new biotypes of the planthopper.

The use of insecticides is not always an efficient means of control, because the population is present throughout the year. Insecticides can help to reduce vector populations in temperate regions if applications are timed to coincide with major migrations of the planthopper into a given area.

Selected References

Chen, C. C., and Chiu, R. J. 1982. Three symptomatologic types of rice virus diseases related to grassy stunt in Taiwan. Plant Dis. 66:15-18.

Hibino, H. 1986. Rice grassy stunt virus. Descriptions of Plant Viruses, No. 320. Commonwealth Mycological Institute and Association of Applied Biologists, Kew, England.

Hibino, H., Cabauatan, P. Q., Omura, T., and Tsuchizaki, T. 1985. Rice grassy stunt virus strain causing tungrolike symptoms in the Philippines. Plant Dis. 69:538-541.

Hibino, H., Usugi, T., Omura, T., Tsuchizaki, T., Shohara, K., and Iwasaki, M. 1985. Rice grassy stunt virus: A planthopper-borne circular filament. Phytopathology 75:894-899.

(Prepared by H. Hibino)

Hoja Blanca

Hoja blanca occurs in Central and South America and the Caribbean and is the only recognized virus disease of rice on the American continents. It was observed in Colombia as early as 1935 but did not become an economically important disease until 1956–1965, when it was responsible for large yield losses in several countries. From 1967 to 1980 the disease incidence was low, but during 1981–1985 the disease became a problem again, especially in Central America. Hoja blanca was present on rice in the southern United States during 1957–1959, but programs established to eradicate the planthopper vector and the virus were successful, and the disease has not reoccurred there.

Symptoms

Infected plants are stunted, and the leaves have chlorotic or yellow stripes or are mottled (Plates 95–97). In some cases, the whole leaf blade may turn white. The panicles of infected plants are often incompletely exserted, and flowers are sterile or absent. Also, the palea and lemma are distorted and turn brown. Infected plants also have fewer and smaller roots, and the roots may be necrotic. Plants infected at the seedling stage may be killed. Symptoms develop from 5 to 34 days after infection, depending on the age of the plant, the cultivar, and the site of infection.

Causal Agent

Rice hoja blanca virus (RHBV) is likely a member of the tenuivirus (rice stripe virus) group. The virus particles are filamentous, 3–4 nm wide, with a spiral or helical configuration, and of undetermined length. The genome consists of single-stranded RNAs, which are likely of both positive and negative senses. RHBV particles have one capsid protein, and a noncapsid protein is produced in large amounts in infected plants. RHBV is closely related serologically to Echinochloa hoja blanca virus (EHBV), which produces symptoms like RHBV on the weed host *Echinochloa colonum* (L.) Link and is transmitted by the vector *Sogatodes cubanus* (Crawford). EHBV may be a strain of RHBV (see EHBV under Minor Virus Diseases).

Vectors

Sogatodes orizicola (Muir) is the primary vector of RHBV. *S. cubanus* is capable of transmitting RHBV from rice to other grasses such as *Echinochloa* but not from rice to rice or from grasses to rice. Also, *S. cubanus* does not colonize rice. RHBV is circulative, propagative, and passed transovarially in the vectors. As much as 60–100% of the progeny from infective females of *S. orizicola* are congenitally infected with RHBV. *S. orizicola* can acquire the virus in a 1-hr acquisition access period (minimum 15 min), and after a long latent period of 30–36 days, can transmit the virus after an inoculation access period of 30 min to 1 hr. The latent period of RHBV exceeds the average life span of the male vectors (14–24 days) and approaches or exceeds the average life span of the female vectors (30–44 days). Planthoppers that acquire the virus through feeding on infected plants, therefore, rarely live long enough to transmit it, and RHBV is transmitted primarily by individuals that acquire the virus transovarially. Field populations of *S. orizicola* generally have 5–15% active transmitters.

RBHV has a deleterious effect on the planthopper vectors. Viruliferous planthoppers have a shortened life span, low fecundity, and a reduced number of viable eggs.

Disease Cycle and Epidemiology

RHBV persists in rice and in the planthopper vectors. Incidence of RHBV is dependent primarily on the number of congenitally infective progeny that can transmit the virus. Infected rice plants probably do not serve as a source of inoculum for spread of the disease during a single season, because the latent period of RHBV is so long in *S. orizicola*. The cyclical appearance of the disease may be explained by the deleterious effect of RHBV on the vector population.

Control

Many cultivars are resistant to RHBV or the vector and have been used extensively. Also, planting rice at a time when the vector population is low can effectively reduce disease incidence.

Selected References

Falk, B. W., Morales, F. J., Tsai, J. H., and Niessen, A. I. 1987. Serological and biochemical properties of the capsid and major noncapsid proteins of maize stripe, rice hoja blanca, and Echinochloa hoja blanca viruses. Phytopathology 77:196-201.

Galvez, E. G. E. 1967. Hoja blanca disease of rice. Pages 35-49 in: The Virus Diseases of the Rice Plant. Johns Hopkins University Press, Baltimore, for the International Rice Research Institute, Los Baños, Philippines.

Galvez, E. G. E. 1967. Transmission of hoja blanca virus of rice. Pages 155-163 in: The Virus Diseases of the Rice Plant. Johns Hopkins University Press, Baltimore, for the International Rice Research Institute, Los Baños, Philippines.

Jennings, P. R., and Pineda T., A. 1971. The effect of the hoja blanca virus on its insect vector. Phytopathology 61:142-143.

Morales, F. J., and Niessen, A. I. 1985. Rice hoja blanca virus. Descriptions of Plant Viruses, No. 299. Commonwealth Mycological Institute and Association of Applied Biologists, Kew, England.

(Prepared by H. Hibino)

Necrosis Mosaic

Necrosis mosaic occurs in central and southern Japan and has been reported from India but not confirmed to be present there. The disease has not been economically important to date.

Symptoms

Infected plants have elongate or spindle-shaped, yellow flecks and streaks on the lower leaves and, sometimes, necrotic flecks on the basal portions of stems and sheaths (Plate 98). Infected plants also are moderately stunted, have a reduced number of tillers, and have a spreading growth habit.

Causal Agent

Rice necrosis mosaic virus (RNMV) is similar to members of the potyvirus group but is not aphid transmitted. A new group, bymovirus, has been proposed for RNMV and other related viruses. RNMV particles are flexuous rods, 13–14 nm in width, and have modal lengths of 205 and 550 nm. RNMV is serologically related to barley yellow mosaic and wheat yellow mosaic viruses. In infected rice cells, RNMV is present as scattered particles in the cytoplasm. Pinwheel and membranous inclusion bodies are formed in virus-infected cells. Some RNMV isolates are readily transmitted to rice by mechanical means; others can be transmitted mechanically with difficulty. The virus is not seed transmitted as was once reported.

Vector

A soilborne fungus, *Polymyxa graminis* Ledingham, is considered to be the vector of RNMV. Resting spores of *P. graminis* are frequently found in the roots of infected rice plants. Also, rice plants become infected with RNMV when grown in sterilized soil to which resting spores of *P. graminis* from infected plants have been added. Rice seedlings are infected with the virus in a 1-day incubation in infective soil.

Disease Cycle and Epidemiology

RNMV persists in soil, rice, and a few weed hosts. Soil retains infectivity for many years. Infection is favored by soil temperatures from 25 to 30°C and low soil moisture. Because flooded soil is not favorable for infection, RNMV is rare in rice seeded directly into flooded fields and occurs primarily in upland rice or in seedling nurseries that are not flooded. The role of weeds in the disease cycle is not known.

Control

In Japan, rice seedlings are generally infected in the seedbeds before they are transplanted to the field. The use of flooded seedbeds or direct seeding into flooded paddies reduces RNMV incidence. Also, treatment of seedbeds with fungicides or chemical sterilants that eliminate the putative vector effectively reduce RNMV.

Selected References

Ghosh, S. K. 1980. Rice necrosis mosaic. Proc. Indian Acad. Sci. Plant Sci. 89:291-299.

Inoue, T., and Fujii, S. 1977. Rice necrosis mosaic virus. Descriptions of Plant Viruses, No. 172. Commonwealth Mycological Institute and Association of Applied Biologists, Kew, England.

(Prepared by H. Hibino)

Ragged Stunt

Ragged stunt was first recognized in 1976 in Indonesia. The disease has been reported from Thailand, Malaysia, India, Sri Lanka, China, Taiwan, Japan, and the Philippines. Ragged stunt epidemics occurred in Indonesia and the Philippines in 1977–1981 and in Thailand in 1980–1982.

Symptoms

Plants infected with the virus are stunted and develop short, dark green leaves that are serrated along one or both edges, giving them a ragged appearance (Plate 99). The upper portions of the leaf blades are often twisted, forming a spiral. Leaf serrations and twisting of the leaves are most pronounced during the early growth stages and on the flag leaf. Vein swellings or galls, formed by hyperplasia of the phloem parenchyma cells, appear on the underside of the leaf blade and on the outer surface of the leaf sheath near the collar. The galls are pale yellow to sometimes brown and up to 10 × 1 mm. Panicle emergence is delayed and incomplete, and the panicles are short and contain unfilled kernels. Branches that form small panicles may grow from the upper nodes of infected plants. Symptoms develop 2–3 weeks after infection. Rice ragged stunt virus is similar to rice black-streaked dwarf virus and rice gall dwarf virus in symptomatology.

Causal Agent

Rice ragged stunt virus (RRSV) is most similar to the fijivirus group, but at present RRSV is unclassified. In common with the fijiviruses, the genome of RRSV consists of 10 double-stranded RNA segments. In contrast to the fijiviruses, RRSV does not appear to have a double shell, and the B spikes of RRSV appear broader at the base. RRSV (mature) particles are generally 63–65 nm in diameter and are composed of five proteins (Fig. 9). In the plant host, virus particles are restricted to the phloem parenchyma and sieve cells, where they occur free in cytoplasm or in viroplasms in the cytoplasm.

Vectors

The primary vector of RRSV is the brown planthopper, *Nilaparvata lugens* (Stål), which is an important pest of rice itself. The virus is also transmitted by *N. bakeri* (Muir). RRSV is circulative and propagative in the insect vectors. *N. lugens* acquires RRSV from infected plants in a 1-day acquisition access period. The latent period of RRSV is 3–35 days, with an average latent period of 8–15 days. The vectors transmit RRSV in a 6-hr inoculation access period (minimum 1 hr). Viruliferous insects remain infective for life. Planthoppers that are infective lay fewer eggs. The percentage of active transmitters in populations of *N. lugens* ranges from 3 to 48% and is 5% in *N. bakeri.*

Disease Cycle and Epidemiology

RRSV persists in rice and in the brown planthopper. Natural infection of the wild rice species *Oryza nivara* Sharma & Shastry and *O. latifolia* Desv. has been reported. The brown planthopper is nearly monophagous on rice, and, consequently, natural infection of weeds and other cereals is rare or nonexistent. RRSV incidence and density of the brown planthopper are high in regions of the tropics where rice is grown year-round. Migratory flights of the planthopper from the tropics are responsible for dispersing RRSV to summer-grown rice in temperate areas.

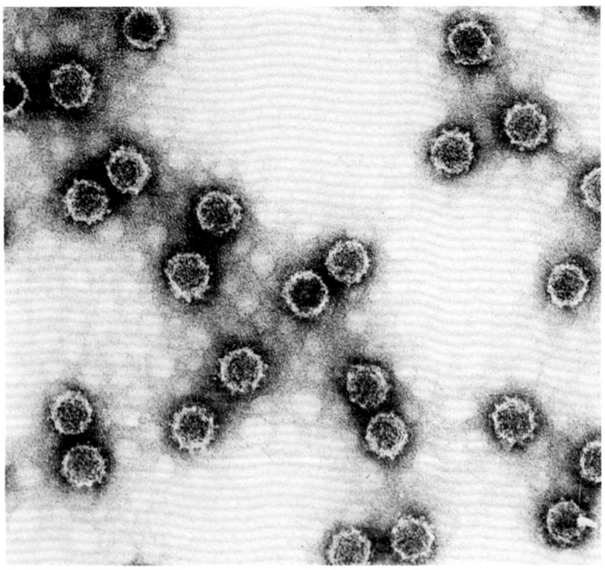

Fig. 9. Rice ragged stunt virus particles from a purified preparation (×100,000).

Control

Rice cultivars are available that are resistant to the brown planthopper and/or resistant to RRSV. Vector-resistant cultivars have low RRSV incidence. In the tropics, insecticides to control the planthopper are not always efficient. In temperate regions, disease incidence may be reduced by insecticides that lower the population of migratory brown planthoppers.

Selected References

Hagiwara, K., Minobe, Y., Nozu, Y., Hibino, H., Kimura, I., and Omura, T. 1986. Component proteins and structure of rice ragged stunt virus. J. Gen. Virol. 67:1711-1715.

Hibino, H., and Kimura, I. 1982. Detection of rice ragged stunt virus in insect vectors by enzyme-linked immunosorbent assay. Phytopathology 72:656-659.

Milne, R. G., Boccardo, G., and Ling, K. C. 1982. Rice ragged stunt virus. Descriptions of Plant Viruses, No. 248. Commonwealth Mycological Institute and Association of Applied Biologists, Kew, England.

Omura, T., Minobe, Y., Kimura, I., Hibino, H., Tsuchizaki, T., and Saito, Y., 1983. Improved purification procedure and RNA segments of rice ragged stunt virus. Ann. Phytopathol. Soc. Jpn. 49:670-675.

(Prepared by H. Hibino)

Stripe

Rice stripe was first recognized in Japan in 1890. It also occurs in China, Korea, Taiwan, and the Soviet Union.

Symptoms

Plants infected at an early stage develop chlorotic leaves, which remain folded and wilt and droop as they elongate. Also, plants infected early are considerably stunted and may die prematurely. Plants infected at later growth stages develop leaves with chlorotic stripes or mottling and necrotic streaks (Plates 100 and 101). Infected plants produce fewer tillers, and panicles are poorly developed or absent. Panicles that are produced have malformed spikelets, and panicle exsertion is often incomplete. Symptoms develop 10–25 days after infection, depending on the age of the rice plant.

Causal Agent

Rice stripe virus (RSV) is the type member of the tenuivirus group, which also includes maize stripe virus (MStpV), rice hoja blanca virus (RHBV), and possibly rice grassy stunt virus (RGSV). Virus particles are circular filaments, 8 nm in width and 290, 510, 610, 840, or 2,110 nm in length. The genome consists of five single-stranded RNA segments of positive and negative senses. The virus sediments as five components, M1, M2, M, B, nB, and a top component. Probably only nB is infective. RSV is closely related serologically to MStpV and is distantly related serologically to RGSV. RSV, MStpV, and RHBV are similar in particle morphology and production of large amounts of a noncapsid protein (molecular weight 16,000) in infected plants. In plant host cells, the virus occurs in masses in the cytoplasm and nucleus.

Vectors

The major vector of RSV is *Laodelphax striatellus* (Fallén). RSV is also transmitted by *Unkanodes sapporona*

(Matsumura), *U. albifascia* Matsumura, and *Terthron albovittatus* (Matsumura). RSV is circulative and propagative and is passed transovarially. The planthoppers acquire the virus in a 1-day acquisition access period (minimum 10 min). The latent period of RSV is 4–30 days (average 8–10 days). Planthoppers transmit the virus after an inoculation access period of 1 hr (minimum 3 min). The percentage of transmitters in vector populations exposed to RSV varies from 14 to 54%. Transovarial passage of RSV occurs at a high rate, and for *L. striatellus*, 42–100% of the progeny from infective females are congenitally infected. RSV can be passed transovarially for at least 40 generations over a period of 6 years. Congenitally infected progeny can transmit RSV immediately after hatching and are very efficient transmitters of the virus throughout the nymphal stages and during the first 2–3 weeks of the adult stage.

Infective planthoppers have a shortened life span, high nymphal mortality, and lower fecundity; females produce fewer eggs, and fewer eggs are viable.

Disease Cycle and Epidemiology

RSV persists in rice, wheat, barley, oats, foxtail millet, rye, wild grasses, and in the vector planthoppers. Maize can also be naturally infected with RSV, but the planthopper vectors do not colonize maize. In Japan, *L. striatellus* overwinters as fourth-stage diapausing nymphs on weeds and on winter cereals, preferentially wheat and barley. In early summer, the first-generation adults move from the winter grasses to rice and transmit RSV. Incidence of RSV is greater in rice-growing regions with large amounts of winter wheat and barley that can support a large overwintering vector population. In central Japan, wheat is the most important source of vectors, because it is harvested late and is present when rice is transplanted. Incidence of RSV is also highest in rice planted early, since the vector density is greatest in the first adult generation. RSV can be spread over long distances by migrations of the small brown planthopper.

Control

Planting cultivars with resistance to RSV and applying insecticides to reduce the vector density of the first-generation adults decrease the incidence of RSV. Also, timing planting to avoid the dispersal of the first-generation planthoppers from overwintering sites would be effective in reducing RSV incidence.

Selected References

Gingery, R. E., Nault, L. R., and Yamashita, S. 1983. Relationship between maize stripe virus and rice stripe virus. J. Gen. Virol. 64:1765-1770.

Ishikawa, K., Omura, T., and Hibino, H. 1989. Morphological characteristics of rice stripe viruses. J. Gen. Virol. 70:3465-3468.

Kisimoto, R., and Yamada, Y. 1986. A planthopper-rice virus epidemiology model: Rice stripe and small brown planthopper, *Laodelphax striatellus* Fallén. Pages 327-344 in: Plant Virus Epidemics: Monitoring, Modelling and Predicting Outbreaks. G. D. McClean, R. G. Garrett, and W. G. Ruesink, eds. Academic Press Australia, Sydney.

Toriyama, S. 1982. Characterization of rice stripe virus: A heavy component carrying infectivity. J. Gen. Virol. 61:187-195.

Toriyama, S. 1983. Rice stripe virus. Descriptions of Plant Viruses, No. 269. Commonwealth Mycological Institute and Association of Applied Biologists, Kew, England.

(Prepared by H. Hibino)

Stripe Necrosis

Stripe necrosis was first reported in 1977 from the Ivory Coast and has not been reported elsewhere.

Symptoms

Infected plants are stunted and have a reduced number of tillers. Leaves of infected plants have chlorotic stripes and exhibit necrosis.

Causal Agent

Rice stripe necrosis virus (RSNV) is possibly a member of the furovirus group. Particles are rod-shaped, 20 nm in width and of two lengths, 110–160 nm or 270–380 nm. RSNV is mechanically transmissible from rice sap to *Chenopodium amaranticolor* Coste & A. Reynier, where it produces chlorotic lesions on the leaves, and to *Nicotiana benthamiana* Domin, where it causes necrotic ringspots on the leaves. Mechanical transmission of RSNV to rice seedlings has not been successful.

Vector

The virus is soilborne, and evidence suggests that it is vectored by the fungus *Polymyxa graminis* Ledingham. Roots of infected plants contain resting spores of the fungus.

Selected Reference

Fauquet, C., and Thouvenel, J. C. 1983. Presence of a new rod-shaped virus in rice plants with stripe necrotic symptoms in Ivory Coast. (In French; English summary.) C. R. Seances Acad. Sci., Ser. 3 296:575-580.

(Prepared by H. Hibino)

Transitory Yellowing

Transitory yellowing occurs in China, Japan, Taiwan, and Thailand and was epidemic during the 1960s and 1970s in China and Taiwan.

Symptoms

Infected rice plants are stunted and have a reduced number of tillers, and the lower leaves turn yellow to bright yellow or orange (Plate 102). The yellowing of the leaves usually starts at the tip and progresses downward (Plate 103). Plants that exhibit leaf discoloration during early growth may produce normal leaves during subsequent growth stages and appear to recover, and hence the name *transitory yellowing*. At

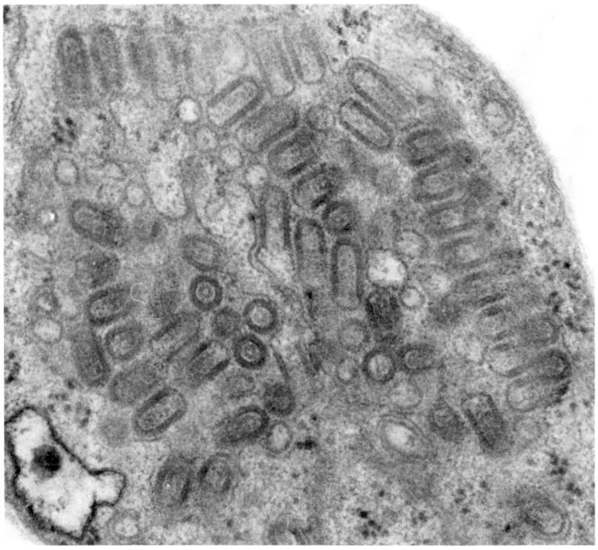

Fig. 10. A thin section of a phloem parenchyma cell in a rice plant infected with transitory yellowing virus, showing oblique and cross sections of virus particles (×45,000). (Courtesy T. Omura)

heading, infected plants are noticeable because they produce no panicles or poorly formed panicles. Symptoms develop 2–4 weeks after infection. Starch accumulates in the leaves of infected plants, as with tungro infection.

Causal Agent

Rice transitory yellowing virus (RTYV) belongs to the plant rhabdovirus group. Virus particles are bullet-shaped, 180–210 × 94 nm, and consist of a lipid-protein envelope with four proteins surrounding RNA (Fig. 10). In infected rice plants, RTYV is restricted to the parenchyma cells bordering the phloem and the sieve elements, where it occurs in the perinuclear space.

Vectors

RTYV is transmitted by *Nephotettix cincticeps* (Uhler), *N. nigropictus* (Stål), and *N. virescens* (Distant). The virus is circulative and propagative in the vectors. Leafhoppers acquire RTYV in a 1-day acquisition access period (minimum 5–10 min). After a latent period of 4–34 days (average 9–16 days), the leafhoppers can transmit RTYV in an 8-hr inoculation access period (minimum 5–10 min). Viruliferous leafhoppers remain infective until death. RTYV causes an increase in the nymphal period of *N. cincticeps* and a reduction in the number of eggs laid.

Disease Cycle and Epidemiology

RTYV persists in rice, a few weeds, and the vector leafhoppers. The importance of alternate hosts as a reservoir for RTYV is not known. Leafhopper populations during the winter have a low percentage of infective individuals. In the early spring, the leafhoppers acquire RTYV from ratoons of infected rice stubble. In Taiwan, the disease incidence is higher in the second rice crop when an infective vector population has built up. Also, in Taiwan RTYV incidence is greater after warm winters that favor the survival of infected rice stubble.

Control

Cultivars with resistance to RTYV are available. Also, application of insecticides in the seedbed or during the early growth stages to control the vector leafhoppers effectively reduces RTYV incidence.

Selected References

Chiu, R.-J., Hsu, Y.-H., Chen, M.-J., Chen, C.-C., Lee, R. C.-R., Lin, M.-C., Lin, S.-M., and Kuo, T.-T. 1990. Purification and partial characterization of rice transitory yellowing virus. Phytopathology 80:777-783.

Chiu, R.-J., Jean, J.-H., Chen, M.-H., and Lo, T.-C. 1968. Transmission of transitory yellowing virus of rice by two leafhoppers. Phytopathology 58:740-745.

Hayashi, T., and Minobe, Y. 1985. Protein composition of rice transitory yellowing virus. Microbiol. Immunol. 29:169-172.

Shikata, E. 1972. Rice transitory yellowing virus. Descriptions of Plant Viruses, No. 100. Commonwealth Mycological Institute and Association of Applied Biologists, Kew, England.

Xie, L. H., and Lin, J. Y. 1980. Studies on the epidemic forecasting of rice transitory yellowing and rice dwarf disease. J. Fujian Agric. Coll. 2:32-43.

(Prepared by H. Hibino)

Tungro and Waika

Tungro is one of the most destructive diseases of rice in South and Southeast Asia, where epidemics of the disease have occurred since the mid-1960s. In the Philippine dialect Ilocano, *tungro* means degenerated growth (Plate 104). Tungro is known by various names in different countries: *mentek* or *penyakit habang* in Indonesia, *penyakit merah* in Malaysia, and yellow-orange leaf in Thailand. Tungro is caused by a dual infection of rice tungro bacilliform virus (RTBV) and rice tungro spherical virus (RTSV). During 1972–1973, an outbreak of a rice virus disease called *waika*, meaning stunting, occurred in southern Japan. Rice waika virus (RWV) is very closely related or identical to RTSV.

Symptoms

Typical symptoms of rice infected with both RTBV and RTSV are stunting and yellow or yellow-orange discoloration of the leaves (Plate 105). Infected plants also have a reduced number of tillers and may have rust-colored spots on the leaves. The degree of stunting and leaf discoloration is dependent on the rice cultivar, strains of the viruses, age of the plant when infected, and the environment. Tungro may produce a mottling, striping, or interveinal necrosis of the leaves. In cultivars that have some resistance to the disease, infected plants exhibit no discoloration or mild discoloration that may disappear as the plants mature. Panicle exsertion of infected plants is delayed and often incomplete, and panicles are short and produce sterile or partially filled grains. Starch accumulates in the leaves of tungro-infected rice plants.

Plants infected with RTBV alone develop symptoms that are similar to but milder than those caused by simultaneous infection by both RTBV and RTSV. Plants infected with RTSV alone may be symptomless or exhibit only a mild stunting. Plants infected with RTSV or RWV at the seedling stage showed 10–40% reduction in yield.

Causal Agents

RTBV and RTSV are both unclassified at the present time. RTBV particles are bacilliform, measure 100–300 × 30–35 nm, and are composed of one major protein (Fig. 11). The genome consists of a double-stranded, circular DNA. RTSV particles are isometric, measure 30 nm in diameter, are composed of two major proteins, and contain a single-stranded RNA (Fig. 12). RTBV and RTSV multiply independently.

In the plant host, both RTBV and RTSV are restricted to the phloem cells and occur free in the cytoplasm. RTSV particles may also occur in viroplasmlike inclusions in the cytoplasm.

Vectors

Both RTBV and RTSV are transmitted by *Nephotettix virescens* (Distant), *N. cincticeps* (Uhler), *N. nigropictus* (Stål),

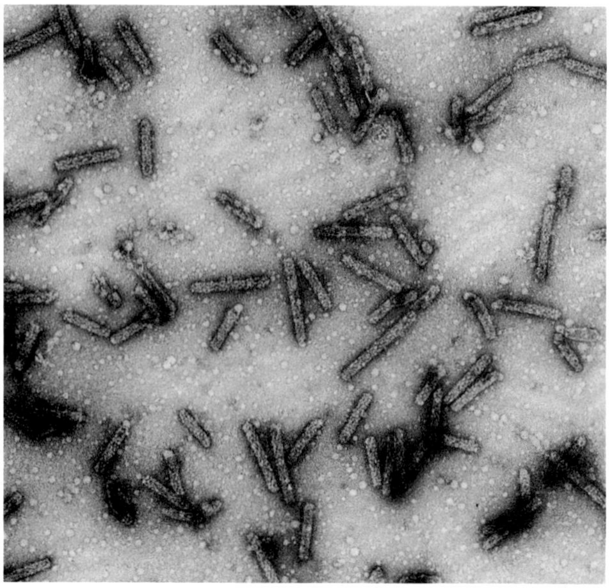

Fig. 11. Rice tungro bacilliform virus particles from a purified preparation (×50,000).

N. parvus Ishihara, *N. malayanus* Ishihara & Kawase, and *Recilia dorsalis* (Motschulsky), and the viruses can be acquired and transmitted simultaneously. RTSV can be acquired independently of RTBV, but acquisition of RTBV is dependent on RTSV.

Leafhoppers do not acquire RTBV by feeding on plants infected with RTBV alone unless they have fed previously on RTSV-infected plants. Transmission of the viruses is dependent on the vector species, vector colonies, and the rice cultivar. Generally, RTBV and RTSV together or RTSV alone are transmitted with high efficiency by *N. virescens*, with intermediate efficiency by *N. cincticeps*, and with low efficiency by *N. nigropictus*, *N. malayanus*, and *R. dorsalis*. Leafhoppers infected with both viruses that feed on cultivars with vector resistance transmit primarily RTBV alone.

RTBV and RTSV are noncirculative and nonpropagative. Both viruses are retained for 3–5 days by *N. virescens*, and for 1 day by *N. nigropictus*, *N. malayanus*, and *R. dorsalis*. *N. virescens* exposed to doubly infected plants acquires the viruses in an 8-hr acquisition access period (minimum 30 min) and can then immediately transmit both viruses together or either virus alone during a 1-hr inoculation access period (minimum 7 min).

Disease Cycle and Epidemiology

Both RTBV and RTSV persist in rice, several weed hosts, and probably in wild rice species. Ratoons from overwintered, infected rice stubble serve as reservoirs of the viruses. The importance of alternate hosts in the disease cycle of tungro has not been determined. Disease incidence and severity depend on the rice cultivar, time of infection, leafhopper species present, population of the vector species, time of planting, and intensity of rice cultivation. Tungro incidence is generally highest in areas where rice is cropped continuously.

Control

Rice cultivars resistant to the vectors of tungro, or less commonly to RTSV, have been developed and used widely for disease control. Cultivars with resistance to both RTBV and RTSV have also been identified. Following intensive cultivation, the vector resistance of some cultivars has broken down under the development of virulent leafhopper populations. In Indonesia, tungro has been successfully managed by scheduled planting to obtain synchrony in rice development and by the rotation of vector-resistant cultivars. Insecticides to lower the vector populations have been effective with moderately resistant cultivars but have not always been successful in controlling tungro on susceptible cultivars. Applications of the antifeedant neem to paddy soils reduced tungro incidence to levels comparable to that obtained with carbofuran in experiments in the Philippines. Other experiments in the Philippines showed that the use of insecticides solely on a border around a rice field controlled tungro as well as if the entire field were treated with insecticide, thereby reducing control costs and minimizing damage to beneficial insects.

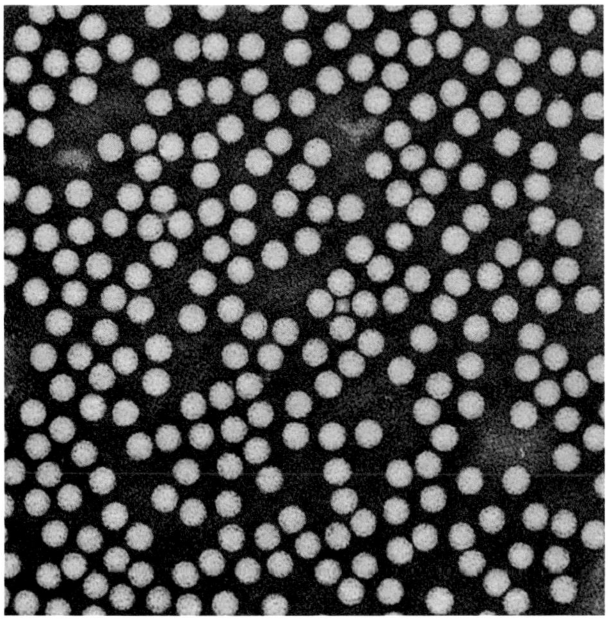

Fig. 12. Rice tungro spherical virus particles from a purified preparation (×120,000).

Selected References

Bajet, N. B., Aguiero, V. M., Daquioag, R. D., Jonson, G. B., Cabunagan, R. C., Mesina, E. M., and Hibino, H. 1986. Occurrence and spread of rice tungro spherical virus in the Philippines. Plant Dis. 70:971-973.

Hibino, H. 1983. Transmission of two rice tungro-associated viruses and rice waika virus from doubly or singly infected source plants by leafhopper vectors. Plant Dis. 67:774-777.

Hibino, H., and Cabauatan, P. Q. 1987. Infectivity neutralization of rice tungro-associated viruses acquired by vector leafhoppers. Phytopathology 77:473-476.

Hibino, H., Roechan, M., and Sudavisman, S. 1978. Association of two types of virus particles with penyakit habang (tungro disease) of rice in Indonesia. Phytopathology 68:1412-1416.

Hibino, H., Tiongco, E. R., Cabunagan, R. C., and Flores, Z. M. 1987. Resistance to rice tungro-associated viruses in rice under experimental natural conditions. Phytopathology 77:871-875.

Omura, T., Saito, Y., Usugi, T., and Hibino, H. 1983. Purification and serology of rice tungro spherical and rice tungro bacilliform viruses. Ann. Phytopathol. Soc. Jpn. 49:73-76.

Saxena, R. C., Justo, H. D., Jr., and Palanginan, E. L. 1988. Trap crop for *Nephotettix virescens* (Homoptera: Cicadellidae) and tungro management in rice. J. Econ. Entomol. 81:1485-1488.

Saxena, R. C., Kahn, Z. R., and Bajet, N. B. 1987. Reduction of tungro virus transmission by *Nephotettix virescens* (Homoptera: Cicadellidae) in neem cake–treated rice seedlings. J. Econ. Entomol. 80:1079-1082.

(Prepared by H. Hibino)

Yellow Mottle

Yellow mottle occurs on the African continent in Ghana, Burkina Faso, Ivory Coast, Kenya, Liberia, Niger, Nigeria, Mali, Sierra Leone, Guinea, Tanzania, and Upper Volta.

Symptoms

Infected plants are stunted and produce fewer tillers. Leaves are mottled or have yellowish streaks and are often twisted or crinkled (Plates 106 and 107). Panicles of infected plants show incomplete emergence and sterile kernels (Plate 108).

Causal Agent

Rice yellow mottle virus (RYMV) is most similar to viruses in the sobemovirus group but is presently unclassified. Virus particles are isometric, about 30 nm in diameter, and the genome consists of a single strand of RNA. RYMV is stable in vitro and can be easily transmitted mechanically.

RYMV occurs in high concentrations in the cytoplasm of the mesophyll and epidermal cells of infected rice plants. The virus particles are scattered or aggregated into crystalline arrays. Virus-specific fibrils and tubules are also present in the cytoplasm of infected cells. The virus can also be found in guttation drops from the leaves.

Vectors

RYMV is transmitted by adult chrysomelid beetles: *Sesselia pusilla* Gerstaecker, *Chaetocnema* spp., *Apophylia* spp., and *Trichispa sericea* Guerin. *Chaetocnema* species are abundant in rice fields. RYMV is noncirculative in the beetle vectors. The beetles acquire RYMV in a 15-min access feeding and can immediately transmit it in a 15- to 30-min inoculation feeding. The beetles usually retain RYMV for 2–3 days (up to 5–8 days).

Disease Cycle and Epidemiology

RYMV persists in rice, *Oryza longistaminata* A. Chev. & Roehr., and several other grass weed species. RYMV may have been endemic in *O. longistaminata* before cultivation of *O. sativa* L. was introduced. The disease usually first occurs along the edges of rice fields. RYMV may be mechanically transmitted from plant to plant when leaves and roots are trimmed during transplanting operations. The virus has been reported in the paddy water of heavily infected fields.

Control

Many upland rice cultivars, including traditional African upland cultivars, are tolerant of RYMV and develop only mild symptoms when infected (Plate 109).

Selected References

Bakker, W. 1970. Rice yellow mottle, a mechanically transmissible virus disease of rice in Kenya. Neth. J. Plant Pathol. 76:53-63.

Bakker, W. 1974. Characterization and ecological aspects of rice yellow mottle virus in Kenya. Agric. Res. Rep. (Wageningen) No. 829.

Bakker, W. 1975. Rice yellow mottle virus. Descriptions of Plant Viruses, No. 149. Commonwealth Mycological Institute and Association of Applied Biologists, Kew, England.

(Prepared by H. Hibino)

Minor Virus Diseases

Rice Tarsonemid Mite Virus

Disease caused by rice tarsonemid mite virus (RTMV) is associated with a sheath browning and grain sterility syndrome of rice in Japan and the Philippines and is transmitted by the tarsonemid mites *Steneotarsonemus spinki* Smiley or by *S. madecassus* Smiley. The virus particles are orbicular, 35 × 16 nm, and the genome consists of two single-stranded RNA segments. RTMV is not systemic and only occurs in the epidermal and mesophyll cells of the leaf sheaths and hulls of the kernels where the mites are present. RTMV is probably a mite virus that can also reproduce in plant cells.

Viruses Associated with Downy Mildew

Two morphologically distinct types of virus particles are found in plants infected with downy mildew. The viruses occur inside the mycelium and oospores of the fungus *Sclerophthora macrospora* (Sacc.) Thirumalachar, C. G. Shaw, & Narasimhan. The viruses are isometric; one is 32 nm and the other 35 nm in diameter. The capsid of the smaller virus is composed of two proteins, and the genome consists of three single-stranded RNA segments. The larger virus has one capsid protein and a genome consisting of one single strand of RNA.

Echinochloa Hoja Blanca Virus

Echinochloa hoja blanca virus (EHBV) can be transmitted to rice from jungle ricegrass, *Echinochloa colonum* (L.) Link, by the planthopper vector *Sogatodes cubanus* (Crawford). *S. cubanus* can also transmit rice hoja blanca virus (RHBV) to a limited extent (see Hoja Blanca). EHBV is transmitted to rice with difficulty, however, and *S. cubanus* does not colonize rice. EHBV is very closely related serologically to RHBV and is found readily in areas affected with RHBV.

Echinochloa Ragged Stunt Virus

Echinochloa ragged stunt virus (ERSV) occurs in Taiwan and can be transmitted from *E. colonum* to rice by the plant-hopper *Sogatella longifurcifera* Esaki & Ishihara. The virus is transmitted by the brown planthopper, *Nilaparvata lugens*, and is serologically related to rice ragged stunt virus, which it resembles in symptomatology, particle morphology, and relations to plant cells.

Other Minor Viruses

Several cereal viruses have been transmitted to rice mechanically under experimental conditions, including barley stripe, brome mosaic, maize dwarf mosaic, ryegrass mosaic, sugarcane mosaic, wheat streak mosaic, and potato yellow dwarf viruses.

Selected References

Chen, C. C., Chen, M. J., and Chin, R. J. 1986. Echinochloa ragged stunt: Symptomatology, host range, and transmission. Plant Prot. Bull. 28:371-381.

Falk, B. W., Morales, F. J., Tsai, J. H., and Niessen, A. I. 1987. Serological and biochemical properties of the capsid and major noncapsid proteins of maize stripe, rice hoja blanca, and Echinochloa hoja blanca viruses. Phytopathology 77:196-201.

Honkura, R., Shirako, Y., Ehara, Y. and Yamanaka, S. 1983. Two types of virus-like particles isolated from downy mildew diseased rice plants. Ann. Phytopathol. Soc. Jpn. 49:653-658.

Ling, K. C. 1972. Rice Virus Diseases. International Rice Research Institute, Los Baños, Philippines.

Shikata, E., Kawano, S., Senboku, T., Tiongco, E. R., and Mayejima, K. 1984. Small virus-like particles isolated from the leaf sheath tissues of rice plants and from the rice tarsonemid mites *Steneotarsonemid spinki* Smiley (Acarina, Tarsonemidae). Ann. Phytopathol. Soc. Jpn. 5:368-374.

(Prepared by H. Hibino)

Diseases Caused by Mycoplasmalike Organisms

Orange Leaf

Orange leaf has been reported in southern China, Indonesia, Malaysia, Sri Lanka, Thailand, and the Philippines. Orange leaf is sometimes confused with tungro, because the two diseases have somewhat similar symptoms and a common vector.

Symptoms

Symptoms generally develop 13–15 days after infection. Initial symptoms of infected seedlings are short, malformed leaves. The edges of the leaves are serrated and chlorotic, and the leaf tips are twisted. As the infection progresses, the leaves turn yellow-orange, beginning at the tip. Eventually the leaves become entirely orange, roll inward, and dry out. Certain cultivars and plants infected during later growth stages do not develop ragged leaves. Most plants die 35–60 days after infection.

Causal Agent

Before mycoplasmalike organisms (MLOs) were found to be the cause of the disease, orange leaf was thought to be a virus disease because of the symptomatology and vector relationship. The orange leaf MLO is pleomorphic, 50–100 nm in diameter, and bounded by a unit membrane; it contains ribosome particles and DNA fibrils (Fig. 13). The agent is restricted to the phloem sieve tube cells in infected rice plants.

Vector

The zig-zag leafhopper, *Recilia dorsalis* (Motschulsky), is the only known vector of the orange leaf mycoplasma agent. The agent is circulative and propagative in the leafhopper but is not passed transovarially. *R. dorsalis* acquires the agent in a 12-hr acquisition access period and becomes infective after a latent period of 15–33 days. The leafhopper can transmit the agent in a 2-hr inoculation access period.

Disease Cycle

The orange leaf MLO persists in rice and the vector leafhopper. Natural infection of other hosts is not known.

Control

No control practices are recommended because the disease is not economically important.

Selected References

Hibino, H., Jonson, G. B., and Sta. Cruz, F. C. 1987. Association of mycoplasmalike organisms with rice orange leaf in the Philippines. Plant Dis. 71:792-794.

Saito, Y., Chaimongkol, Y., Singh, K. G., and Hino, T. 1976. Mycoplasma-like bodies associated with rice orange leaf disease. Plant Dis. Reptr. 60:649-651.

(Prepared by H. Hibino)

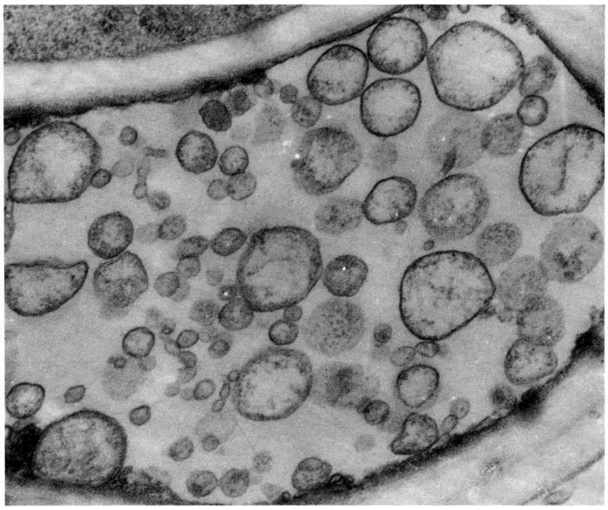

Fig. 13. A thin section of a sieve tube in a plant infected with orange leaf, showing mycoplasmalike organisms (×25,000).

Yellow Dwarf

Yellow dwarf was first recognized in southern Japan about 1910. At present, the disease is widely distributed in rice growing areas in Asia, including India, Bangladesh, Indonesia, Malaysia, Taiwan, Japan, China, and the Philippines. The disease causes economic damage in Japan, Taiwan, and India, but its occurrence elsewhere is not considered to be a serious problem.

Symptoms

Infected plants are stunted, tiller profusely, and show a general chlorosis, producing pale green or yellow leaves (Plate 110). Plants infected at the seedling stage develop symptoms 40–50 days after infection and often die prematurely. Plants that survive produce no panicles or a few defective panicles. Plants infected when they are 60 days old or older develop mild symptoms if any. Ratoons from late-infected plants, however, exhibit definitive symptoms of yellow dwarf.

Causal Agent

Mycoplasmalike organisms (MLOs) were first demonstrated to be associated with the yellow dwarf disease in 1967. The yellow dwarf MLO is pleomorphic, 100–800 nm in diameter, and bounded by a unit membrane; it contains ribosomes and DNA fibrils. The agent is restricted to the phloem sieve tube cells in the rice host.

Vectors

Nephotettix cincticeps (Uhler) is the primary vector of yellow dwarf in temperate regions, and *N. nigropictus* (Stål) and *N. virescens* (Distant) are the major vectors in the tropics. Yellow dwarf is also vectored by *N. parvus* Ishihara and *N. malayanus* Ishihara & Kawase. The MLO is circulative and propagative in the vectors. The leafhoppers acquire the MLO in a 1-hr acquisition access period (minimum 10–30 min) and, after a long latent period of 20–40 days (average 25–30), can transmit the MLO in a 1-hr inoculation access period (minimum 2–10 min). The leafhoppers retain the MLO until death. The percentage of active transmitters in populations of the leafhoppers ranges from 50 to 95%.

Disease Cycle and Epidemiology

Yellow dwarf persists primarily in rice and the leafhopper vectors. Experimental infection of several wild *Oryza* spp. and a few grass weed species has been reported, but alternate hosts are not considered to be an important reservoir of yellow dwarf inoculum. In temperate areas, leafhopper nymphs acquire the MLO from infected rice ratoons after harvest, and overwinter on weeds or winter cereals after the ratoons die. The leafhoppers then transmit the MLO to rice in the spring. In these areas, yellow dwarf incidence is correlated with the number of infective leafhoppers in the overwintering population. In warmer tropical and subtropical areas, the vector may survive on rice and ratoon crops year-round.

Control

In temperate areas in Japan where yellow dwarf is a problem, late planting of rice is recommended to avoid the migration of the vectors from the overwintering hosts, when their population is greatest. In warm climates where two crops or more of rice are grown a year, rogueing of infected plants in the first rice crop helps to reduce disease in the second crop. Applications of insecticides to reduce vector density in the nurseries and fields may also help reduce yellow dwarf incidence. Commonly planted cultivars are susceptible to yellow dwarf, but sources of resistance to the disease and the vectors have been identified and are being used in breeding programs.

Selected References

Ishii, M., Yasuo, S., and Ono, K. 1969. Epidemiological studies of rice yellow dwarf in Kanto-Tosan district, Japan. (In Japanese; English summary.) J. Cent. Agric. Exp. Stn. 13:1-21.

Muniyappa, V., and Raju, B. C. 1981. Response of cultivars and wild species of rice to yellow dwarf disease. Plant Dis. 65:679-680.

Muniyappa, V., and Raychaudhuri, S. P. 1988. Rice yellow dwarf disease. Pages 233-284 in: Mycoplasma Diseases of Crops. K. Maramorosch and S. P. Raychaudhuri, eds. Springer-Verlag, New York.

Shikata, E., and Maramorosch, K. 1967. Presumptive mycoplasma etiology of yellow disease. FAO Plant Prot. Bull. 17:121-128.

(Prepared by H. Hibino)

Diseases Caused by Viruslike Agents

Chlorotic Streak

Chlorotic streak was first reported in India in 1978 and has not been reported elsewhere since. Symptoms are stunting, chlorotic streaking or mottling of young leaves, and poor emergence of panicles with sterile or discolored spikelets. Other symptoms that may occur are leaf distortion, vein swelling, excess tillering, and formation of aerial roots. The symptoms vary considerably depending on the time of infection and the cultivar. The disease is associated with the rice mealy bug, *Brevennia rehi* (*Ripersia oryzae* Green), and reported to be transmitted in a nonpersistent manner. Seedlings infested with mealy bugs that had fed on symptomatic rice plants for 8 hr developed chlorotic streak after 1 or 2 weeks. Plants infested with mealy bugs, but not chlorotic streak, showed stunting and leaf yellowing only. The chlorotic streak syndrome has not been associated with mealybug infestation outside India.

Selected Reference

Anjaneyulu, A., Singh, S. K., Shukua, V. D., and Shenoi, N. M. 1980. Chlorotic streak, a new disease of rice. Curr. Sci. 49:520-521.

(Prepared by H. Hibino)

Crinkle

Crinkle was first observed in patches in upland rice in Sierra Leone in 1976. Since then, it has been observed at a low incidence in the Ivory Coast, central Nigeria, and Liberia.

Leaves of emerging seedlings are crinkled or distorted and may develop pale yellow, lens-shaped lesions. Entire leaves may turn yellow, and roots are stubby and bushy. Seedlings may die or produce new tillers that grow normally. Panicles produced from seedlings with crinkle are poorly filled.

Attempts to transmit the crinkle condition have failed, but the condition was reproduced at a low frequency on healthy plants grown in soil from around diseased plants.

Selected Reference

Ou, S. H. 1985. Rice Diseases. 2nd ed. Commonwealth Mycological Institute, Kew, England.

(Prepared by H. Hibino)

Wrinkled Stunt and Witches'-Broom

Both wrinkled stunt and witches'-broom are associated with particular cultivars of rice, and the conditions are seed-transmitted. Wrinkled stunt occurs in the progeny of the cross IR454 and 71101/9. Symptoms include severe stunting with dark green, wrinkled, twisted leaves. Panicle exsertion is incomplete, and only a few of the spikelets set seed.

Witches'-broom occurs in the progeny of CICA 4. Plants show pronounced stunting and tiller profusely, producing up to 100 tillers. Panicles are produced but are small, with only 10–20 seeds.

Selected Reference

Ou, S. H. 1985. Rice Diseases. 2nd ed. Commonwealth Mycological Institute, Kew, England.

(Prepared by H. Hibino)

Diseases Caused by Nematodes

Nematodes are unsegmented roundworms that are aquatic. Plant-parasitic nematodes are typically microscopic, transparent, and vermiform; exceptions are the females of some genera, which become swollen and saclike. Plant-parasitic nematodes feed and reproduce on living plants and are capable of active migration in the rhizosphere, on aerial plant parts, and/or inside the plant. They are disseminated by wind, irrigation and flood water, tools, machinery, animals, and infected plant propagation materials.

The life cycle of a typical plant-parasitic nematode is simple and direct. An egg laid by the female develops into a first-stage juvenile, which usually undergoes a first molt within the egg shell and hatches as a second-stage juvenile. After hatching, the juvenile develops through two more stages, each completed by a molt, and becomes an adult after the fourth molt. The time required to complete the life cycle varies considerably, from a week to several months, depending on the genus and on temperature and moisture.

More than 150 species of plant-parasitic nematodes have been associated with rice or rice soils. A parasitic relationship with rice, however, has been demonstrated for only a few of these nematode species. Those of economic importance on paddy and deep-water rices are *Aphelenchoides besseyi* Christie, *Ditylenchus angustus* (Butler) Filipjev, *Hirschmanniella* spp., *Criconemella* spp., and *Meloidogyne* spp. Those of importance on upland rice are *Heterodera* spp., *Meloidogyne* spp., *Pratylenchus* spp., and *Hoplolaimus indicus* Das.

Estimates of annual yield losses due to plant-parasitic nematodes on rice worldwide range from 10 to 25%. The lower estimate corresponds to an annual monetary value of US$16 billion. Actual losses due to nematodes may be greater, because nematodes generally do not cause diagnostic symptoms on aboveground parts of plants, and damage is often attributed to other causes, such as nutrient deficiencies or other pathogens.

Selected References

Fortuner, R., and Merny, G. 1979. Root-parasitic nematodes of rice. Rev. Nematol. 2:79-102.

Gerber, K., and Smart, G. C., Jr. 1987. Plant-parasitic nematodes associated with aquatic vascular plants. Pages 488-491 in: Vistas on Nematology: A Commemoration of the Twenty-Fifth Anniversary of the Society of Nematologists. J. A. Veech and D. W. Dickson, eds. Society of Nematologists, Hyattsville, MD.

Hollis, J. P., Jr., and Keoboonrueng, S. 1984. Nematode parasites of rice. Pages 95-146 in: Plant and Insect Nematodes. W. R. Nickle, ed. Marcel Dekker, New York.

(Prepared by J. C. Prot)

White Tip

White tip, caused by *Aphelenchoides besseyi* Christie, has been reported from most of the rice-growing countries of Africa, from North, Central, and South America, and from Asia, Eastern Europe, and the Pacific.

Symptoms

The common name *white tip* is based on the characteristic whitening of 3–5 cm of the leaf tip (Plates 111 and 112). The whitened tips eventually become necrotic and shred. The central part and the base of infected leaves are sometimes darker green than normal. The upper leaves are the most affected, and the flag leaf is often twisted, hindering the emergence of the panicle. Other symptoms include reduction in the length of the panicle, reduction in the number of grains, sterile flowers, misshapen grains, stunting of the plant, late ripening and maturation, and production of tillers from the upper nodes. Not all plants infected with *A. besseyi* show the characteristic leaf and panicle symptoms, although their yields are reduced. Symptoms similar to white tip may result from insect injuries or may be caused by other nematodes, such as *Ditylenchus angustus* (Butler) Filipjev.

Yield losses vary with cultivar, year, temperature, cultural practices, and other variable factors. In infested fields, the average losses range from 10 to 30%. In fields where all plants have been attacked, maximum losses of 70% for the most susceptible cultivars and 20% for the most resistant cultivars have been reported.

Disease Cycle and Epidemiology

A. besseyi parasitizes flooded rice as well as upland rice. The nematode survives from one season to the next in infected seeds and to a lesser extent on weeds and on ratoons and debris from the preceding rice crop. *A. besseyi* can survive in seeds in a dehydrated state for up to 3 years.

Reviviscent nematodes that emerge from seeds after planting are attracted by and feed on the primordia of emerging plants. Later, they move to the growing points of stems and leaves and feed ectoparasitically, causing obstruction of the xylem vessels and disintegration of the phloem cells. The nematode multiplies abundantly and is carried upward as the plant grows. *A. besseyi* eventually migrates to the developing panicle, punctures the inflorescence, and penetrates into the florets via the apicular opening, where it feeds on the ovaries, stamens, and embryos. The rate of reproduction is high in the florets before anthesis. After anthesis, reproduction declines and then ceases in maturing grains, and the nematode becomes dormant as the grains dehydrate.

A. besseyi is inactive if the atmospheric humidity is less than 70%, and the nematode only migrates on aerial plant parts when they are covered with a film of water. The optimum temperature for development of *A. besseyi* is 23–32°C, but the nematode is active from 13 to 43°C. *A. besseyi* is killed by temperatures exceeding 43°C. Ammonium sulfate, ammonium nitrate, calcium superphosphate, and potassium chloride increase the severity of the disease. The nematode is effectively spread by irrigation water.

A. besseyi is usually amphimictic, but parthenogenetic reproduction may occur. There are several generations in one season. The life cycle of the nematode takes 7–15 days, depending on temperature and humidity, and there are therefore several generations during a growing season.

A. besseyi has a wide host range, including Chinese cabbage, chrysanthemum, Italian millet, onion, soybean, strawberry, sugarcane, sweet corn, sweet potato, and yam. This nematode can also feed and complete its life cycle on many species of fungi.

Green weights of some cultivars inoculated simultaneously with *A. besseyi* and the fungus *Sclerotium oryzae* Cattaneo are significantly lower than those of plants inoculated with either organism alone.

Control

Since the nematode is seedborne, hot-water treatment of seed is the best control method for *A. besseyi*. For small quantities of seeds, a hot-water treatment of 10–15 min at 55–61°C, without presoaking, destroys the nematodes without affecting germination. For large quantities of seeds, presoaking for 24 hr in cool water followed by 15 min in water at 51–53°C is recommended. Chemical treatment of seeds with nematicides is also effective.

Burning straw, weeds, and new growth, and seeding directly into standing water are also recommended for control of *A. besseyi*.

Cultivars of rice differ in susceptibility to *A. besseyi*, and several have been reported as resistant: Arkrose; Asahi; Bluebonnet; Bluebonnet 50; Century 52; Century Patna 231; Chinoor; Fortuna; Gumartia; Improved Bluebonnet; Nira; Norin Nos. 6, 8, 37, 39 and 43; Nukata Ginbozu; Rescora; Texas Patna; Toxan 38; and TP49.

Nematicides applied as a root dip before transplanting or to the paddy water during transplanting can also be used to control *A. besseyi*.

Selected References

Fortuner, R., and Orton Williams, K. J. 1975. Review of the literature on *Aphelenchoides besseyi* Christie, 1942, the nematode causing "white tip" disease in rice. Helminthol. Abstr. Ser. B 44:1-14.

Franklin, M. T., and Siddiqi, M. R. 1972. *Aphelenchoides besseyi*. Descriptions of Plant-Parasitic Nematodes, Set 1, No. 4. Commonwealth Institute of Helminthology, St. Albans, England.

McGawley, E. C., Rush, M. C., and Hollis, J. P. 1984. Occurrence of *Aphelenchoides besseyi* in Louisiana rice seed and its interaction with *Sclerotium oryzae* in selected cultivars. J. Nematol. 16:65-68.

Ou, S. H. 1985. Rice Diseases. 2nd ed. Commonwealth Mycological Institute, Kew, England.

Todd, E. H., and Atkins, J. G. 1959. White tip disease of rice. II. Seed treatment studies. Phytopathology 49:184-188.

(Prepared by J. C. Prot)

Ufra

Ufra, caused by *Ditylenchus angustus* (Butler) Filipjev, has been reported from Bangladesh, Burma, Egypt, India (Uttar Pradesh and Ganges Delta), Madagascar, Malaya, the Philippines, southern Thailand, and Vietnam (Mekong Delta).

Symptoms

The symptoms vary considerably with the degree and time of infection. In the field, the earliest symptoms appear on plants that are more than 2 months old. Plants show malformations and chlorosis or mosaic discoloration of the upper leaves (Plates 113 and 114). Later, scattered dark stains appear on the leaves, and parts of the upper internodes of the stem turn dark brown (Plate 115). The disease is most typical and evident on the panicle, where two types of symptoms can be observed: the swollen ufra and the ripe ufra. Swollen ufra occurs when the nematode damages the panicle at an early stage; the panicle coils and remains enclosed within the leaf sheath, and there is a strong tendency for the stem to branch (Plate 116). The ripe ufra occurs when the nematode damages the panicle at a later stage; the panicle emerges but is distorted, bears sterile or empty, distorted spikelets at the base, and produces normal grains only near the tip (Plate 117).

Losses in infested fields range from 20 to 90%, with average losses estimated at 30% (Plate 118).

Disease Cycle and Epidemiology

D. angustus is mainly associated with deep-water rice. However, with the development of intensive rice cropping and irrigation, it has been reported in irrigated rice adjacent to deep-water rice fields.

D. angustus is an obligate ectoparasite. Reproduction is amphimictic, and at least three generations occur in one growing season. The nematode is able to survive from one crop to the next by remaining coiled in rice stubble and debris and in soil. Infested ratoons are also sources of inoculum. *D. angustus* can survive desiccation for at least 6 months, but the number of survivors declines over time, with an average half-life of about 2 weeks. The nematode becomes active when the fields are flooded.

Initially, the nematode parasitizes the terminal buds of newly sprouted seeds. It then migrates upwards as the rice plant grows, feeding on new tissue. During flowering and heading, *D. angustus* is found mainly on the stem just above the nodes, on the peduncles, and inside the glumes. The nematode becomes inactive when the plant matures.

The most favorable temperature for infection of rice is 28–30° C. The nematode is disseminated primarily by irrigation water.

With the exception of *Leersia hexandra* Sw., the known host range of *D. angustus* is confined to *Oryza* spp., namely, *O. alta* Swallen, *O. glaberrima* Steudel, *O. eichingeri* A. Peter, *O. latifolia* Desv., *O. meyerriana* Baill., *O. minuta* J. S. Presl. ex C. B. Presl., and *O. officinalis* Wall ex Watt, in addition to *O. sativa* L.

Control

Lengthening the time between crops by delaying sowing, transplanting after flooding, and using early-maturing cultivars all decrease the severity of ufra. Also, burning crop debris reduces the disease incidence by reducing the primary inoculum. Application of zinc also decreases the severity of the disease on zinc-deficient plants.

Cultivars B-69-1; R-16-06; Habiganj Boro II, IV, and VI; Lalchiken; Lohagor; and Monolata have been reported as resistant to *D. angustus*. In a recent trial conducted in Bangladesh, 116 of 432 entries tested, including Boishbish, Chamara, Kartiskail, Khama, and Shada Pankaish, appeared to be resistant.

Soil treatments with nematicides are effective against the nematode but costly. Nematicidal compounds applied as foliar sprays or at the preflowering stage have been reported to be effective in controlling ufra.

Selected References

Cox, P. G., and Rahman, L. 1979. The overwinter decay of *Ditylenchus angustus*. Int. Rice Res. Newsl. 4:14.

McGeachie, I., and Rahman, L. 1983. Ufra disease: A review and a new approach to control. Trop. Pest Manage. 29:325-332.

Ou, S. H. 1985. Rice Diseases. 2nd ed. Commonwealth Mycological Institute, Kew, England.

Seshadri, A. R., and Dasgupta, D. R. 1975. *Ditylenchus angustus*. Descriptions of Plant-Parasitic Nematodes, Set 5, No. 64. Commonwealth Institute of Helminthology, St. Albans, England.

(Prepared by J. C. Prot)

Rice-Root Nematodes

The common name *rice-root nematode* designates nematodes belonging to the genus *Hirschmanniella*. Several species of *Hirschmanniella* are associated with rice, and at least five may be of economic significance: *H. caudacrena* Sher, *H. imamuri* Sher, *H. mucronata* (Das) Luc & Goodey, *H. oryzae* (van Breda de Haan) Luc & Goodey, and *H. spinicaudata* (Schuurmans-Stekhoven) Luc & Goodey.

H. oryzae is present in most of the rice-growing countries in West Africa, and in Central and South America, Asia, Madagascar, the United States, and Portugal. *H. caudacrena* has been reported from the United States and Vietnam; *H. imamuri* from Japan, Korea, Nigeria, and Vietnam; *H. mucronata* from Bangladesh, India, the Philippines, and Thailand; and *H. spinicaudata* from Cameroon, Cuba, The Gambia, Ivory Coast, Nigeria, Senegal, Venezuela, and Vietnam.

Symptoms

Aboveground symptoms include retarded growth, reduction in the tiller number, stunting, chlorosis, and late maturation in severe cases. These symptoms are not diagnostic, however, because other factors may induce similar symptoms (Plate 119). Histopathological changes in infected roots include cell necrosis and formation of cavities resulting from the disruption of cell walls. It is estimated that 80 million hectares are infested with *Hirschmanniella* spp. Yield reduction estimations range from 10 to 35% in Asia and West Africa.

Disease Cycle and Epidemiology

Hirschmanniella spp. are well adapted to flooding and are more common in wet paddy soils than in well-drained ones. Between rice crops, the nematodes survive on weeds, in dead rice roots, and in soil (Plate 120). *H. oryzae* can survive under dry conditions for at least 5 months and under paddy water for up to 7 months.

Hirschmanniella spp. are migratory endoparasites of roots, but they are sometimes found in the coleoptile. At sowing or transplanting, they enter the seedling roots anywhere from the tip to the base, and all stages can penetrate the roots. The nematodes move within the roots, and a few days after infection, the females lay their eggs. Juveniles hatch and develop in the root cortex. Reproduction in *Hirschmanniella* spp. is bisexual. *H. imamuri* has one generation per crop, but *H. oryzae* may have two generations per crop. The nematodes generally leave the roots when the panicle appears, but many remain inside the roots until harvest.

Additional hosts for rice-root nematodes include many rice field weeds. Cabbage, cotton, maize, and sugarcane have also been reported to be hosts for *H. oryzae*.

Control

Incorporation of straw ash, neem, and mustard cakes into paddy soil is reportedly effective in reducing the population of rice-root nematodes. Intercropping with trap plants such as *Sesbania rostrata* Brem. or *Sphenoclea zeylanica* Gaertn., which are toxic to *Hirschmanniella* spp., also can be used to control rice-root nematodes. Biological agents such as nematode-trapping fungi and sulfate-reducing bacteria have also been reported to be effective against *Hirschmanniella* spp.

Some cultivars show a high level of resistance to rice-root nematodes, containing less than one nematode per gram of roots. In comparison, susceptible cultivars may have more than 100 nematodes per gram of roots. Only the cultivar TKMg has been reported to be immune to rice-root nematodes.

Nematicides applied as soil treatments or as root dips prior to transplanting have been reported to reduce rice-root nematode incidence in soil and seedling roots and significantly increase yields.

Selected References

Fortuner, R. 1985. Nematode damage assessment problems and solution for a rice nematode: *Hirschmanniella oryzae*. Calif. Plant Pest Dis. Rep. 4:2-7.

Hollis, J. P., Jr., and Keoboonrueng, S. 1984. Nematode parasites of rice. Pages 95-146 in: Plant and Insect Nematodes. W. R. Nickle, ed. Marcel Dekker, New York.

Luc, M., and Fortuner, R. 1975. *Hirschmanniella spinicaudata*. Descriptions of Plant-Parasitic Nematodes, Set 5, No. 68. Commonwealth Institute of Helminthology, St. Albans, England.

Ou, S. H. 1985. Rice Diseases. 2nd ed. Commonwealth Mycological Institute, Kew, England.

Pariselle, A. 1987. Comportement du nematode *Hirschmanniella oryzae* (van Breda de Haan) dans les racines de la legumineuse *Sesbania rostrata* Brem. Rev. Nematol. 10:333-336.

Prasad, J. S., Panwar, M. S., and Roa, Y. S. 1986. Chemical control of the root nematode (*Hirschmanniella mucronata*) in rice. Beitr. Trop. Landwirts. Veterinarmed. 24:65-69.

Prasad, J. S., and Rao, Y. S. 1984. The status of root nematodes (*Hirschmanniella* spp.) as pests of rice (*Oryza sativa*). Beitr. Trop. Landwirts. Veterinarmed. 22:281-284.

Siddiqi, M. R. 1973. *Hirschmanniella oryzae*. Descriptions of Plant-Parasitic Nematodes, Set 2, No. 26. Commonwealth Institute of Helminthology, St. Albans, England.

(Prepared by J. C. Prot)

Root-Knot Nematodes

Several species of *Meloidogyne* are associated with rice: *M. arenaria* (Neal) Chitwood, *M. incognita* (Kofoid & White) Chitwood, *M. javanica* (Treub.) Chitwood, *M. graminicola* Golden & Birchfield, *M. oryzae* Mass, Sanders, & Dede, and *M. salasi* Lopez.

M. arenaria, *M. incognita*, and *M. javanica* are ubiquitous and of major economic significance in temperate, warm, and tropical regions. *M. graminicola* has been reported from Bangladesh, India, Laos, Thailand, the United States, and Vietnam; *M. oryzae* from Surinam; and *M. salasi* from Costa Rica and Panama.

Symptoms

Aboveground symptoms associated with the presence of *Meloidogyne* spp. include chlorosis, wilting, retarded maturation, and reduction in growth and tillering. *M. graminicola* reduces the elongation ability of deep-water varieties.

Root-knot nematodes induce hypertrophy and hyperplasia in the root meristem, cortex, endodermis, and xylem and the formation of giant cells in the meristem, cortex, and xylem. These transformations result in the formation of characteristic root galls and prevent root elongation (Plate 121).

On upland rice, root-knot nematodes can cause substantial yield loss. Average yield losses of 10–20% (up to 50%) have been observed in infested upland rice in West Africa and Asia. On irrigated or flooded rice, losses due to root-knot nematodes have not been estimated under field conditions. However, loss due to *M. graminicola* may be important in deep-water rice, because most plants infected with large numbers of the nematode are unable to grow above the water.

Disease Cycle and Epidemiology

Root-knot nematodes are poorly adapted to flooded conditions. *M. arenaria*, *M. incognita*, and *M. javanica* do not survive continuous flooding. *M. graminicola* does not invade rice roots growing in flooded soils, but it is able to survive inside roots for at least 5 weeks and in flooded soil for 5 months or more. Between rice crops, the nematodes survive as eggs in the soil or continue to reproduce on other crops or on weeds. The largest populations of *Meloidogyne* spp. are found in sandy or coarse-textured soils, and in these soils they may be found up to a depth of 30 cm. Clay soils are less favorable for *Meloidogyne* spp., and the nematodes are confined to the top 10 cm. Soil temperatures that are optimum for invasion of the roots and development and reproduction of the nematodes range from 25 to 30° C.

Root-knot nematodes are obligate endoparasites. The typical life cycle is exemplified by *M. graminicola*. Second-stage juveniles of *M. graminicola* infect the apical meristem of young seedling roots before flooding. Glandular secretions released during feeding cause the development of giant cells and root galls within 4 days. The nematode develops and reproduces by meiotic parthenogenesis within the host tissues. The females lay their eggs inside the galls, and juveniles hatch from the eggs and reinfect the same root. Inside the roots, *M. graminicola* can develop and reproduce normally for at least 36 days. In well-drained soil, at 22–29° C, the life cycle of *M. graminicola* is completed in 19 days. Nitrogen and phosphorus increase rate of development of *M. graminicola* progeny.

M. incognita and *M. javanica* can infect more than 700 plant species, and *M. arenaria* more than 300. Recorded hosts include numerous vegetables and food crops, cereals, cash crops, fruit trees, grasses, pasture legumes, and ornamentals. *M. graminicola* has a wide host range, including oats, bushbean, tomato, potato, plantain, sorghum, several species of the Cyperaceae and Poaceae, and *Sphenoclea zeylanica*.

Simultaneous infection by *M. javanica* and *Pyricularia grisea* (Cooke) Sacc. reduces the growth of rice plants more than either pathogen alone. Synergistic interactions between *M. javanica* and *Fusarium* spp. have also been reported.

Control

Continuous flooding is very effective for control of *M. arenaria*, *M. incognita*, and *M. javanica*, and it reduces infestation by *M. graminicola* to a minimum. Fallow or crop rotation with nonhost plants also reduces root-knot nematode populations. Rotations with groundnut, sweet potato, maize, and soybean are effective in reducing *M. graminicola* populations.

Cultivars resistant to *M. graminicola* include Baharsia, Bonnet 73, CH46, Dumai, Garem, Hamsa, IR2, IR5, IR47, LA110, Manoharsali, and Patni 6. Cultivars resistant to *M. incognita* and/or *M. javanica* include IR20, IS335, IS358, International, and Peta. Cultivars of *Oryza glaberrima* Steudel generally appear to be more resistant or tolerant of *M. incognita* and *M. javanica* than cultivars of *O. sativa* L.

Several nematicidal compounds applied as soil drenches, foliar sprays, seedling root dips, or seed soaks or in irrigation water give good control of root-knot nematodes and result in increased yields. Chemical treatments on a field scale are considered too costly, but they may be economical in nurseries.

Selected References

Bridge, J., and Page, S. L. J. 1982. The rice root-knot nematode, *Meloidogyne graminicola* on deep-water rice (*Oryza sativa* subsp. *indica*). Rev. Nematol. 5:225-232.

Diomande, M. 1981. Effets direct et combine des engrais et de *Meloidogyne incognita* sur le riz pluvial. Rev. Nematol. 4:71-74.

Ibrahim, I. A., Ibrahim, I. K. A., and Rezk, M. A. 1973. Host-parasite relationship of *Meloidogyne incognita* (Kofoid and White) Chitw. on rice. Nematol. Mediterr. 1:8-14.

Lopez, R. 1984. *Meloidogyne salasi* sp. nov. (Nematoda: Meloidogynidae), a new parasite of rice (*Oryza sativa* L.) from Costa Rica and Panama. Turrialba 34:275-286.

Maas, P. W. T., Sanders, H., and Dede, J. 1972. *Meloidogyne oryzae* n. sp. (Nematoda: Meloidogynidae) infesting irrigated rice in Surinam (South America). Nematologica 24:305-311.

Mulk, M. M. 1976. *Meloidogyne graminicola*. Descriptions of Plant-Parasitic Nematodes, Set 6, No. 87. Commonwealth Institute of Helminthology, St. Albans, England.

Orton Williams, K. J. 1972. *Meloidogyne javanica*. Descriptions of Plant-Parasitic Nematodes, Set 1, No. 3. Commonwealth Institute of Helminthology, St. Albans, England.

Orton Williams, K. J. 1973. *Meloidogyne incognita*. Descriptions of Plant-Parasitic Nematodes, Set 2, No. 18. Commonwealth Institute of Helminthology, St. Albans, England.

Orton Williams, K. J. 1975. *Meloidogyne arenaria*. Descriptions of Plant-Parasitic Nematodes, Set 5, No. 62. Commonwealth Institute of Helminthology, St. Albans, England.

Ou, S. H. 1985. Rice Diseases. 2nd ed. Commonwealth Mycological Institute, Kew, England.

Prasad, K. S. K., and Rao, Y. S. 1976. Chemotherapy of the root-knot nematode (*Meloidogyne graminicola*) in rice. II. Effects of root-dip treatments with nematicides on invasion and development of the nematode. Z. Pflanzenkr. Pflanzenschutz 83:730-735.

Prasad, K. S. K., and Rao, Y. S. 1984. Effect of foliar application of systemic pesticides on the development of *Meloidogyne graminicola* in rice. Indian J. Nematol. 14:125-127.

Roy, A. K. 1978. Effectiveness of rotation with nonhost or fallow on reducing infestation of *Meloidogyne graminicola*. Indian J. Nematol. 8:156-158.

(Prepared by J. C. Prot)

Cyst Nematodes

Four species of cyst nematodes may cause economic damage to upland rice: *Heterodera elachista* Ohshima, *H. oryzae* Luc & Berdon, *H. oryzicola* Rao & Jayaprakash, and *H. sacchari* Luc & Merny.

H. sacchari has been reported from the Congo, Ivory Coast, Nigeria, Senegal, and Trinidad; *H. oryzae* from Bangladesh, India, Ivory Coast, Senegal, and Japan; *H. elachista* from Japan; and *H. oryzicola* from India.

Symptoms

No diagnostic symptoms are associated with rice cyst nematode infection. In severe cases, plants are stunted and have retarded growth, leaves are severely chlorotic, and the number of tillers is reduced. In Japan, *H. elachista* has been considered responsible for the failure of continuous upland rice cropping. In inoculation experiments, *H. elachista* suppressed yields 7.2–18% and *H. sacchari* suppressed yields 20–75%.

Disease Cycle and Epidemiology

Rice cyst nematodes are mainly associated with upland rice. However, *H. oryzae* and *H. sacchari* have been found on flooded rice, and *H. elachista* has been found on flooded rice sown in unflooded soil.

Between rice crops, *Heterodera* spp. survive as eggs in cysts and egg masses. When rice is present, second-stage juveniles hatch from the eggs and penetrate the rice roots. They affix their heads in the central root cylinder, and the cells there are transformed into syncytia. Juveniles can develop into males or females, and reproduction is amphimictic. Females lay eggs in a gelatinous matrix (the egg mass) or retain the eggs in their body, which turns into a cyst. The life cycle of *Heterodera* spp. takes approximately 1 month, and three generations can infect the same rice crop.

H. sacchari also infects sugarcane; *H. oryzae* is able to reproduce on banana; and *H. elachista* can reproduce on *Echinochloa crus-galli* (L.) P. Beauv.

Reproduction of *H. oryzicola* is inhibited in the presence of *Meloidogyne graminicola* Golden & Birchfield. *H. oryzicola* and the seedling blight fungus *Sclerotium rolfsii* Sacc. cause higher seedling mortality when they infect simultaneously than when either pathogen infects alone.

Control

Continuous flooding, fallow periods, or crop rotation with soybeans or sweet potatoes reduce populations of *H. elachista*.

The cultivars Lalnakanda 41, CR143-2-2, and TKM6 have been reported resistant to *H. oryzicola*.

Soil fumigation significantly reduced populations of *H. elachista* and significantly increased yield.

Selected References

Babatola, J. O. 1983. Pathogenicity of *Heterodera sacchari* on rice. Nematol. Mediterr. 11:21-25.

Jayaprakash, A., and Rao, Y. S. 1982. Life history and behavior of the cyst nematode, *Heterodera oryzicola* Rao and Jayaprakash, 1978 in rice (*Oryza sativa* L.). Proc. Indian Acad. Sci. Anim. Sci. 91:283-295.

Jayaprakash, A., and Rao, Y. S. 1984. Cyst nematode, *Heterodera oryzicola* and seedling blight fungus *Sclerotium rolfsii* disease complex in rice. Indian J. Nematol. 14:58-59.

Luc, M., and Taylor, D. P. 1977. *Heterodera oryzae*. Descriptions of Plant-Parasitic Nematodes, Set 7, No. 91. Commonwealth Institute of Helminthology, St. Albans, England.

Nishizawa, T., Shimizu, K., and Nagashima, T. 1972. Chemical and cultural control of the rice cyst nematode *Heterodera oryzae* Luc et Berdon Brizuela, and hatching responses of the larvae to some root extracts. Jpn. J. Nematol. 2:27-32.

Ohshima, Y. 1974. *Heterodera elachista* n. sp. an upland rice cyst nematode from Japan. Jpn. J. Nematol. 4:51-56.

Rao, Y. S., and Jayaprakash, A. 1978. *Heterodera oryzicola* n. sp. (Nematoda: Heteroderidae) a cyst nematode on rice (*Oryza sativa* L.) from Kerala State, India. Nematologica 24:341-346.

Shimizu, K. 1977. Seasonal fluctuation and rhizospherical distribution of population of the upland rice cyst nematode, *Heterodera elachista*. Jpn. J. Nematol. 7:49-57.

(Prepared by J. C. Prot)

Other Nematodes

Stunt Nematodes

Tylenchorhynchus brassicae Siddiqi, *T. gladiolatus* Fortuner & Amougou, *T. martini* Fielding, and *T. mashhoodi* Siddiqi & Basir are not adapted to continuous flooding but may cause economic damage in rainfed rice. These nematodes suppress the length and dry weight of roots in inoculation experiments.

Ring Nematodes

Criconemella onoensis Luc, which causes characteristic knotting on roots, significantly reduces rice yield in Louisiana and Texas.

Other *Criconemella* spp. have been associated with rice in many rice-growing countries.

Lesion Nematodes

Pratylenchus spp. commonly parasitize roots of upland rice. Severe damage has been reported for *P. zeae* Graham in Zimbabwe and *P. indicus* Das in India.

Lance Nematodes

Hoplolaimus indicus Das parasitizes upland rice in India. Associated with the fungus *Sclerotium rolfsii* Sacc., it increases the severity of seedling blight.

Selected References

Das, P. K., and Rao, Y. S. 1970. Life history and pathogenesis of *Hoplolaimus indicus* incidence in rice. Indian Phytopathol. 23:459-464.

Hollis, J. P., Jr., and Keoboonrueng, S. 1984. Nematode parasites of rice. Pages 95-146 in: Plant and Insect Nematodes. W. R. Nickle, ed. Marcel Dekker, New York.

Joshi, M. M, and Hollis, J. P. 1976. Pathogenicity of *Tylenchorhynchus martini* swarmers to rice. Nematologica 22:123-124.

Khan, E., and Chawla, M. L. 1975. *Hoplolaimus indicus*. Descriptions of Plant-Parasitic Nematodes, Set 5, No. 66. Commonwealth Institute of Helminthology, St. Albans, England.

Prasad, J. S., and Rao, Y. S. 1983. Influence of age of rice plant at the time of inoculation of *Pratylenchus indicus* Das on the multiplication of the nematode and yield of host plant. Acta Oecol. Oecol. Plant. 4:309-312.

Ramana, K. V., Mathur, S. C., and Rao, Y. S. 1974. Role of *Hoplolaimus indicus* on the severity of seedling blight of rice. Curr. Sci. 43:687-688.

(Prepared by J. C. Prot)

Part II. Abiotic Diseases and Disorders

Rice plants may develop abiotic diseases with characteristic symptoms in response to nutrient deficiencies or when they are exposed to toxic concentrations of naturally occurring or synthetic chemicals in the soil and water environment. Stress symptoms in rice resulting from abiotic causes often may be confused with symptoms caused by infectious disease organisms (bacteria, fungi, nematodes, or viruses). The type and severity of injury depend on the plant growth stage and plant part affected by the abiotic disorder as well as the type and concentration of the nutrient or toxin. The ability to distinguish abiotic from biotic diseases and to identify their specific causes is necessary for effective crop management in areas where abiotic diseases are prevalent.

(Prepared by D. M. Brandon)

Akiochi (Hydrogen Sulfide Toxicity)

Akiochi is the most frequent and serious abiotic rice disorder in Japan and California caused by a naturally occurring toxic substance. Akiochi is caused by hydrogen sulfide (H_2S) toxicity, which most frequently occurs in lighter-textured sandy to silty clay loam paddy soils. These soils often have a lower cation exchange capacity and active iron (Fe) content than do heavy clay rice soils in which the problem usually does not occur. Hydrogen sulfide toxicity occurs in soils relatively high in organic matter and soluble sulfates that become highly reduced after flooding. Hydrogen sulfide gradually accumulates in these flooded soils, especially where there is little active Fe to precipitate the sulfide ion. The concentration of free H_2S is usually very low in typical clay rice soils, which are not affected by the problem because they contain active Fe that reacts with H_2S to form insoluble FeS_2 and FeS.

Symptoms of H_2S toxicity first become apparent on the rice plant near the end of tillering and increase in severity as the plant matures (Plate 122). The Japanese word *akiochi* (autumn decline) describes the gradual decline of the rice with maturity. The plant usually shows normal growth and color in early stages of development, with the first symptoms being a change in root color from white or light brown to black (Plate 123). The black root color of affected plants is a surface coating of iron that can be rubbed off in early stages of the toxicity, but later the whole root becomes black and begins to deteriorate. Grayish blotches first appear in irregular patterns along the margins of older leaves and later in younger leaves. Affected leaves begin to roll and gradually turn brown from the tip down. Unaffected areas on the leaves may retain a normal green color. Plants in advanced stages of H_2S toxicity wilt, appear drought stricken, and produce sterile panicles or light grain or both. Symptoms often occur in irregular areas of a field, reflecting differences in soil type, water depth, organic matter content, and other environmental variables. In advanced stages of akiochi, soil with a high H_2S accumulation emits an odor similar to that of rotten eggs.

Control

The best control measure is to drain fields to supply oxygen to affected plants.

Selected References

Baba, I., Inada, K., and Tajima, K. 1965. Mineral nutrition and occurrence of physiological diseases. Pages 173-195 in: The Mineral Nutrition of the Rice Plant. Johns Hopkins University Press, Baltimore.

Jashi, M. M. 1974. Physiological responses of rice to hydrogen sulfide. Ph.D. diss. Louisiana State University, Baton Rouge.

Mitsui, S. 1965. Dynamic aspects of nutrient uptake. Pages 53-62 in: The Mineral Nutrition of the Rice Plant. Johns Hopkins University Press, Baltimore.

Ponnamperuma, F. N. 1972. The chemistry of submerged soils. Adv. Agron. 24:29-96.

Statewide Integrated Pest Management Project. 1983. Integrated Pest Management of Rice. Publ. 3280. University of California, Division of Agricultural Sciences, Berkeley.

Yoshida, S. 1981. Fundamentals of Rice Crop Science. International Rice Research Institute, Los Banos, Philippines.

(Prepared by D. M. Brandon)

Akegare (Organic Acids and Toxic Gases)

Akegare is a disorder that develops in the early stages of rice growth, in contrast to akiochi, which is expressed in the later reproductive and ripening stages. Akegare occurs in many countries of Asia, but its cause is not clearly identified in the literature. It frequently occurs in both light-textured and clay soils that are poorly drained and have a relatively high organic matter content. Descriptions of akegare in Japan indicate that it is similar to a disorder of rice in California caused by accumulation of toxic organic acids and gases released by decomposition of organic matter in flooded soils. The anaerobic decomposition of organic matter can produce toxic organic acids, such as butyric, acetic, and propionic acids, which eventually form toxic gases in the soil, including carbon dioxide and methane. The presence of toxic gases is often detected by bubbles that rise to the surface of the paddy water, especially when the soil is disturbed by stirring. Soil incorporation of large quantities of fresh organic materials, such as green manures and straw, promotes production of organic acids and gases in submerged soils.

Plant symptoms caused by toxic organic acids and gases usually appear early in the season, when the rice plant is in the seedling to early tillering stage. Initial symptoms appear about 20 days after transplanting, or 25 to 35 days after seeding in the direct water-seeded cultural system. Plant growth is slowed; leaves may initially become dark green and then rapidly turn bright yellow. The roots of affected plants are shorter, larger in diameter, and have fewer root hairs than roots of unaffected plants. Roots may become brown as they die, but they do not turn black as do roots affected by H_2S toxicity. Plants may die in large areas of a field where soil organic matter remains high and the problem is severe, or damage may be limited to retarded growth and low grain yield in moderately affected areas.

Control

Two methods used to control akegare are applications of nonsulfate fertilizers and surface drainage at late tillering to supply oxygen. Green manure crops should be avoided where the disorder occurs.

Selected References

Baba, I., Inada, K., and Tajima, K. 1965. Mineral nutrition and occurrence of physiological diseases. Pages 173-195 in: The Mineral Nutrition of the Rice Plant. Johns Hopkins University Press, Baltimore.

Yoshida, S. 1981. Fundamentals of Rice Crop Science. International Rice Research Institute, Los Baños, Philippines.

(Prepared by D. M. Brandon)

Straighthead

Straighthead is the most frequently observed abiotic disorder of rice in North America. It has also been reported in Cuba, Guyana, Mexico, Panama, and Japan. The disorder most frequently occurs in light-textured loam, sandy loam, and clay loam soils of the southern United States, but it has also been observed in rice grown on clay soils. The cause of straighthead is not known exactly, but arsenic and hydrogen sulfide induce it in the southern United States in light-textured soils that are continuously flooded. Organic matter incorporated into the soil and high nitrogen rates in soils where straighthead occurs increase the severity of the disorder. It is most severe when rice follows nonflooded crops or is planted in soils that have not previously produced rice.

Straighthead symptoms and severity on rice are reduced by draining the field to oxygenate the soil at the stage of late tillering to panicle initiation. This indicates that soil anaerobiosis transforms chemical compounds to their reduced states under prolonged flooding, resulting in solubilization and accumulation of toxic substances such as arsenic and hydrogen sulfide in the soil solution. Drainage may alleviate the disorder by reoxidizing the soil, which in turn increases the soil sorption capacity and reduces the concentration of toxic substances in the soil solution. Reduced iron and manganese compounds may release sulfide and arsenic into the soil solution during flooding, and reoxidation of these compounds during drainage may increase the soil sorption capacity and lower their concentration in the soil solution.

The characteristic symptoms of straighthead are not visible until the late boot to heading stages of rice. Plants in affected areas appear vigorous and dark green, with no discernible symptoms until heading. Affected plants may be darker green with more erect leaves than normal plants, but this symptom is not always apparent where the disorder occurs. The most characteristic symptom of straighthead is erect panicles that are partially to completely sterile and of insufficient weight to cause the panicle to tip over during the grain-filling stage as in normal rice (Plate 124). The panicles of severely affected plants may not emerge or may emerge incompletely from the boot. The florets are usually sterile, and the shapes of the lemma and palea are distorted, especially in long-grain varieties. Typically the lemma or palea, or both, are crescent shaped, resembling a parrot beak (Plate 125). Distortion of the floret is not as common in straighthead-tolerant cultivars or short- and medium-grain cultivars. The florets may be absent in severe cases, with the panicle consisting of only branches. Affected plants usually remain green, and their senescence is greatly delayed compared with that of normal rice plants. Failure to form seed results in no translocation of carbohydrates from the leaves and stems, which causes the plants to remain green and produce numerous tillers from lower nodes.

Control

Control is achieved by draining fields at the onset of the plants' reproductive stage. Cultivars with tolerance to straighthead are available in some areas.

Selected References

Atkins, J. G. 1974. Rice Diseases of the Americas, Review of Literature. U.S. Dep. Agric. Agric. Res. Serv. Agric. Handb. 448.

Grist, D. H. 1986. Rice. 6th ed. Longman Group, Harlow, England.

Jashi, M. M., Ibrahim, I. K. A., and Hollis, J. P. 1975. Hydrogen sulfide: Effects on the physiology of rice plants and relationship to straighthead disease. Phytopathology 65:1165-1170.

Statewide Integrated Pest Management Project. 1983. Integrated Pest Management of Rice. Publ. 3280. University of California, Division of Agricultural Sciences, Berkeley.

Yoshida, S. 1981. Fundamentals of Rice Crop Science. International Rice Research Institute, Los Baños, Philippines.

(Prepared by D. M. Brandon)

Nutrient Deficiencies and Toxicities

Nutrient deficiencies and toxicities in rice are common in most rice production areas of the world. Affected plants develop distinctive symptoms, which usually occur in irregular patterns in the field. The irregular patterns may be caused by inherent soil differences, such as sandy streaks, variations in pH or salinity, and variable nutrient content or availability. They are also often associated with uneven fertilizer application, tillage practices, and management of residue of previous crops.

Symptoms of deficiencies and toxicities may be confused with those of infectious diseases. Accurate identification of symptoms requires a knowledge of rice symptomology and complete information about previous crop history, soil treatments, production practices for the current and previous crops, and the time of symptom appearance. Soil and plant analyses in combination with a knowledge of symptoms caused by other rice diseases are often necessary to positively identify the cause of nutrient deficiencies or toxicities.

When possible, moderate symptoms should be used in diagnosis, because they are usually more definitive than symptoms in severely affected plants. Leaves of plants that are severely deficient are usually dead or dying and may not reveal the unique symptoms associated with a particular deficiency.

In some cases, rice plants may not clearly display the symptoms described below. In other cases, the same symptoms may be associated with other disorders. Symptoms associated with nutrient deficiencies or toxicities should be confirmed with plant tissue analysis or soil analysis or both to allow positive diagnosis and proper management of these problems.

Nutrient Deficiencies Causing Color Change in Young Leaves

Leaf symptoms caused by deficiencies of nutrients with immobile ions (Fe, Zn, Mn, B, and Cu) usually occur in young or upper leaves, because immobile ions are not translocated from older to younger leaves. Consequently, older leaves may appear more normal in color and size during the early stages of these deficiencies. This characteristic can be confusing in plants recovering from certain deficiencies. A zinc deficiency may induce mild symptoms of leaf yellowing, but roots of young plants often develop sufficient mass to absorb new sources of zinc, and new growth begins to appear a healthy green, while older leaves remain yellow or brown.

Iron Deficiency

Iron deficiency causes younger leaves to lose their green color and turn pale yellow (chlorotic) to whitish. Older leaves may become chlorotic. Plants with severe deficiency usually die. Initial symptoms usually become visible about 4–5 weeks after seeding.

Manganese Deficiency

Manganese deficiency produces interveinal chlorotic streaks that form from the tip to the base of the leaf, later becoming dark brown and necrotic. Plants are stunted. Newly emerging leaves are short, narrow, and light green.

Zinc Deficiency

In water-seeded, continuously flooded rice with zinc deficiency, the midribs of younger leaves turn yellow, especially at the base, and the chlorosis spreads over the entire leaf. The initial symptoms appear when seedlings are in the four-leaf stage (about 20–25 days after water seeding). Affected leaves tend to be droopy and float on the water rather than grow erect (Plate 126). Brown blotches and streaks later appear on older leaves (Plate 127), plant growth is stunted, and plants may die in cases of severe deficiency. Tillering is reduced and maturity delayed in cases of moderate deficiency. Open water void of rice plants marks the site of severe deficiency in water-seeded rice.

In dry-seeded rice, the characteristic symptom of zinc deficiency is small brownish spots that form on the leaf tip and eventually over the whole leaf surface, giving the field a bronzed appearance, soon after permanent flooding (Plate 128). Older leaves may become bronzed, and they may become necrotic in cases of severe deficiency. New leaves of recovering plants may be green, with older leaves showing advanced symptoms.

Copper Deficiency

Leaves of plants with copper deficiency turn a bluish green and become chlorotic near the tip. The chlorosis spreads along the midrib. Dark brown necrosis develops later near the tip. Newly emerging leaves may fail to unfold and appear needle-like.

Nutrient Deficiencies and Toxicities Causing Color Change in Older Leaves

Symptoms caused by deficiencies of nutrients with mobile ions (such as N, K, P, and S) usually occur initially in the lower leaves, with new leaves showing no symptoms. These mobile nutrients are translocated from older leaves to younger leaves and prevent the development of symptoms in younger leaves, especially in cases of mild deficiency.

Nitrogen Deficiency

Nitrogen deficiency causes older leaves to change from green to light green or yellow; new leaves are darker green. Leaves are narrow, short, and erect (Plate 129), and in advanced stages of deficiency they die, turning straw-colored. Severely affected plants are stunted and produce few tillers.

Phosphorus Deficiency

Leaves of plants with phosophorus deficiency initially turn yellow but then rapidly turn a grayish green. Plants are stunted and produce few tillers. Leaves are narrow, short, erect, rolled, and brittle. Older leaves often die, becoming necrotic in water-seeded rice (Plate 130). Affected plants change to a dark bluish green during the period from late tillering to panicle initiation. Heading (flowering) and maturation are noticeably delayed.

Potassium Deficiency

On plants with potassium deficiency, leaves are short, drooping, and dark green, with chlorosis developing between veins near the leaf tip. Small brown spots develop near the leaf tip and coalesce to form blotches toward the leaf base as the deficiency progresses (Plates 131 and 132). Symptoms are most severe on older leaves, which dry to a light brown. Plants may be slightly stunted, but tillering is only slightly reduced.

Sulfur Deficiency

Leaf symptoms of sulfur deficiency are very similar to those of nitrogen deficiency. Leaves become pale green, and chlorosis develops, more slowly and less intensively than in nitrogen deficiency.

Iron Toxicity

Iron toxicity causes very small brown spots on lower leaves, appearing first on the leaf tip and spreading toward the base. The spots coalesce in interveinal areas. Older leaves may become chlorotic but usually remain green. Entire leaves may appear purplish brown in severe cases.

Boron Toxicity

Boron toxicity causes chlorosis at the tips of older leaves along the margins and, later, large brown elliptical spots. Leaves eventually turn brown and become necrotic. Severely affected plants are stunted and may die.

Salt Toxicity

On plants exposed to excessive levels of salinity, the tips of older leaves become chlorotic and later turn brown to white. Plant growth is stunted and tillering is reduced if the salinity persists. Whitish deposits of salt often accumulate on plants at the waterline and on soil at the edge of the field.

Selected References

Grist, D. H. 1986. Rice. 6th ed. Longman Group, Harlow, England.

Statewide Integrated Pest Management Project. 1983. Integrated Pest Management of Rice. Publ. 3280. University of California, Division of Agricultural Sciences, Berkeley.

Yoshida, S. 1981. Fundamentals of Rice Crop Science. International Rice Research Institute, Los Baños, Philippines.

(Prepared by D. M. Brandon)

Herbicide Injury

Herbicides may injure rice when they are improperly applied to the crop, when they remain as toxic residues from the previous crop, when they drift into rice from application to adjacent crops, when they contaminate irrigation water sources, and when insecticides predispose rice to herbicide injury. Use of the wrong herbicides, applying herbicides unevenly or at excessive rates, or applying them at sensitive stages of rice growth or to susceptible cultivars can injure rice plants enough to stunt or kill them. Symptoms vary according to the kind of herbicide applied, cultivar susceptibility to the herbicide, and soil and environmental factors. High temperatures during or soon after application are often associated with herbicide injury to rice, especially from propanil, chlorophenoxy herbicides, bentezon, and thiocarbamates.

Common symptoms of herbicide injury in rice plants are presented below. Symptoms vary with the cultivar, the severity of the injury, and the growth stage of the plant at the time of herbicide application and observation of the injury.

Chlorophenoxy Herbicides

In injury due to chlorophenoxy herbicides, such as 2,4-D, MCPA, and 2,4,5-T, the tips of older leaves turn chlorotic and become necrotic if the symptoms persist. Root growth is inhibited, and roots appear enlarged or stubby within 2 weeks after herbicide application. Crowns become necrotic inside, and newly emerging roots look like bulbs, with little or no elongation. Phenoxy herbicides applied at the sensitive growth stage (soon after panicle initiation) cause panicles to become twisted and deformed, and they either do not completely emerge from the boot or emerge from the side of the leaf sheath.

Thiocarbamate Herbicides

The thiocarbamates molinate and thiobencarb cause similar injury symptoms, but those of thiobencarb may be more pronounced. Older leaves of injured seedlings (at the two- to four-leaf stage) turn chlorotic and die rapidly. Seedlings become blue-green, deformed, twisted, and brittle, and a slight swelling occurs at the base of the plant. New leaves developing after the injury are green, spatulate, and shorter than older leaves, which often die. Roots remain healthy, with no apparent symptoms. Symptoms of injury from thiocarbamates applied between late tillering and panicle initiation are similar to those of chlorophenoxy injury. Flag leaves of injured plants become chlorotic and may be twisted around the panicle. Panicles are deformed, emerge incompletely from the sheath, and are partially to completely sterile (Plate 133).

Propanil

Chlorosis develops on leaves exposed to propanil at the time of application (primarily older leaves), beginning at the leaf tip and spreading to the base. Leaves rapidly become necrotic and may die. Newly emerged leaves are green and show no symptoms. Cultivars may differ greatly in susceptibility to propanil phytotoxicity. The application of the insecticides carbaryl, malathion, parathion, or carbofuran within 14 days before or after the application of propanil causes much more severe symptoms of propanil injury. The interaction between propanil and these insecticides causes rapid necrosis, and severely affected plants may die (Plate 134).

Sulfonylurea Herbicides

Older leaves of some cultivars turn chlorotic to purplish green and rapidly become necrotic in areas affected by sulfonylurea injury (Plate 135). Younger (upper) leaves may remain green for some time in cases of slight injury. Older and newly emerging leaves and tillers die in severe cases, and new tillers that are twisted and deformed emerge from lower nodes.

Selected References

Grist, D. H. 1986. Rice. 6th ed. Longman Group, Harlow, England.

Kaufman, P. B., and Krafts, A. S. 1956. Responses of the rice plant to different formulations and methods of application of 2,4-D, 2,4,5-T and MCPA. Hilgardia 24(15):411-453.

Statewide Integrated Pest Management Project. 1983. Integrated Pest Management of Rice. Publ. 3280. University of California, Division of Agricultural Sciences, Berkeley.

(Prepared by D. M. Brandon)

Glossary

C—Celsius (°C = (°F − 32) × 5/9)
g—gram (1 g = 0.035 ounce)
hr—hour
km—kilometer (1 km = 0.62 mile)
m—meter (1 m = 39.37 inches)
mg—milligram (1 mg = 0.001 g)
mm—millimeter (1 mm = 0.001 m)
μm—micrometer (1 μm = 10^{-6} m)
nm—nanometer (1 nm = 10^{-9} m)

acervulus (pl. acervuli)—a more or less saucer-shaped, asexual fungal structure (conidioma) bearing conidiogenous cells and conidia
acicular—needle-shaped, slender and pointed
acquisition access period—the duration of feeding by a vector necessary for it to acquire a virus from an infected plant host
adventitious root—a root produced in an unusual position, such as between nodes or at a node, or at an unusual time in development
agar—gelatin extracted from seaweed
alternate host—any of various plant hosts of a given pathogen
amphimixis (adj. amphimictic)—true sexual reproduction
anamorph—the asexual state of a fungus
anastomosis—fusion of hyphal cells of an individual or individuals, involving the cell wall or cytoplasm
antheridium (pl. antheridia)—a male gametangium
anthesis—the growth stage in which flowers are open and pollen is shed
antibody—a protein (immunoglobulins) formed in an animal in reaction to an antigen and capable of combining with that antigen
antifeedant—a substance that deters insects from feeding on a host plant
antiserum—a preparation from animal blood containing a specific antibody
ascocarp—a specialized, often flask-shaped fungal structure containing asci
ascospore—a fungal spore produced within an ascus as the result of meiosis
ascus (pl. asci)—the saclike fungal structure in which ascospores are produced
auricle—the earlike lobe at the base of the leaf in grasses

basidiospore—a fungal spore produced on a basidium as the result of meiosis
basidium (pl. basidia)—a commonly club-shaped fungal cell or organ producing basidiospores
biotype—a subspecific group of individuals characterized by genetic, physiological, or ecological properties differentiating them from other members of the species
biovar—*see* biotype
boot—the popular name for the leaf sheath that surrounds the inflorescence during its early development in grasses

capsid—the protein coat surrounding the nucleic acid of a virus
catenulate—formed in chains
chlamydospore—a thick-walled, asexual fungal spore formed by modification of a hyphal or conidial cell or cells and generally functioning in survival
chlorosis—yellowing or bleaching of normally green plant tissue due to a lack of chlorophyll
circulative virus—a virus that enters and circulates in the hemolymph of its insect vector
clavate—club-shaped, narrowing toward the base
coalesce—to run together, overlap, merge
congenital—existing at birth
conidioma (pl. conidiomata)—a specialized multihyphal fungal structure bearing conidiophores and conidia
conidiophore—a simple or differentiated fungal hypha that produces conidia
conidium (pl. conidia)—an asexual fungal spore
cortex (adj. cortical)—the tissue between the epidermis and the phloem in plant culms and roots
cross-protection—protection of a plant by a mild strain of a virus from a more severe strain of the same virus
crown—the compacted series of nodes from which the culms and roots develop in grasses
culm—the stem of a grass plant
cultivar—a cultivated variety of a plant
cyst—in nematodes of the genus *Heterodera*, the egg-filled body of a dead female
cytoplasm—the contents of a cell other than the nucleus

diapause—a period of delayed development
dictyochlamydospore—a multicelled chlamydospore
dictyospore—a fungal spore divided into cells by both longitudinal and transverse septa
distichous—arranged in two lines
DNA—deoxyribonucleic acid

ectoparasite—a parasitic organism that lives outside its host
endodermis—the specialized layer of cells that separates the root cortex from the vascular tissue
endoparasite—a parasite that lives within its host
epidemic—widespread outbreak of a disease
epidemiology—the study of factors affecting the outbreak, development, and spread of disease
epidermis—the outermost layer of cells on all plant parts
epiphyte (adj. epiphytic)—an organism that lives on a plant surface but does not cause disease
erumpent—breaking through the surface
etiology—the study of the causes of disease

fascicle—a small group or bundle
fecundity (adj. fecund)—the ability to produce offspring, particularly in large numbers
filiform—threadlike
flagellum (pl. flagella)—a whiplike appendage of bacterial cells or zoospores, functioning in locomotion
floret—a small flower; one of the individual flowers that make up a dense inflorescence, such as those in the spikelets of grasses
fusiform—narrowing toward the ends, football shaped

gall—a localized overgrowth
geniculate—bent sharply, like a knee
genome—the genetic material of an organism
glume—one of the two sterile bracts at the base of a grass spikelet
gramineous—belonging to the grass family (Gramineae, now usually called Poaceae)
gram-positive, gram-negative—pertaining to bacteria that retain or do not retain, respectively, the violet dye in Gram's stain
guttation—exudation of drops of water from the stomates or hydathodes of leaves
guttulate—(of fungal spores) containing one or more oillike droplets

histopathological—concerning changes in normal plant structure due to infection by a pathogen
horizontal resistance—partial resistance effective against all races of a pathogen
host—a plant that can be invaded by and serve as a source of nutrients for a parasite

hyaline—colorless, transparent
hyperplasia—an overgrowth due to excessive cell division
hypertrophy—an overgrowth due to excessive cell enlargement
hypha (pl. hyphae)—one of the tubular filaments that constitute a fungal thallus
hyphopodium (pl. hyphopodia)—a usually pigmented hyphal branch of one or two cells, functioning in attachment to and penetration of a host

in vitro—in culture; artificially maintained outside the host
inclusion body—any of various small particles occurring in cells infected with a virus
inoculation access period—the duration of feeding by an insect vector sufficient for it to transmit a virus to its host
inoculum—a pathogen or its parts that can cause infection
internode—the area of the culm between two adjacent nodes

latent period—the time elapsed between the acquisition of a virus by an insect vector and the point at which the vector is capable of transmitting the virus; the time elapsed between inoculation with a pathogen and the expression of disease symptoms in the inoculated host plant
lemma—the lower of the two bracts (the other of which is the palea) enclosing the flower in grasses
lesion—a localized area of discolored, diseased plant tissue
lodge—to fall over or be beaten down

macroconidium (pl. macroconidia)—one of the larger form of conidia of a fungus that also produces microconidia
mechanical transmission—transmission of a plant virus by the transfer of sap from an infected to a healthy plant, usually via a wound
meiosis (adj. meiotic)—the reduction division of nuclei, in which the chromosome number is reduced from diploid to haploid during sexual reproduction
meristem—plant tissue that functions in cell division and differentiation
mesophyll—photosynthetic tissue composed of parenchyma cells between the upper and the lower epidermis of a leaf
microconidium (pl. microconidia)—one of the smaller form of conidia of a fungus that also produces macroconidia
molt—(of nematodes and insects) to shed the old body cuticle during a phase of growth
monoclonal antibodies—a population of homogenous, or identical, antibodies that recognize the same antigenic determinant
monophagous—feeding only on one kind of food
monophialide (adj. monophialidic)—a fungal phialide having only one opening through which conidia (phialospores) are released
monostichous—arranged in one line
mosaic—disease symptom characterized by patches of normal-colored plant tissue intermixed with patches of light green or yellowish tissue, caused by a viral infection
mottle—an irregular pattern of indistinct light and dark green areas of plant tissue, caused by a virus infection
M_r—relative molecular mass
muriform—having both longitudinal and transverse septa
mycelium (pl. mycelia)—the mass of hyphae that make up the thallus, or body, of a fungus
mycotoxin—any of various toxins, harmful to people and animals, produced by certain fungi growing on foodstuffs

neck—the area of the culm between the panicle and the uppermost leaf
necrosis (adj. necrotic)—death and decay of plant tissue, usually accompanied by darkening and discoloration
node—a joint of a culm or rachis; the site of attachment of a leaf or spikelet
nymph—a juvenile insect in a stage of development in which it resembles its adult form

obclavate—club-shaped, narrowing toward the top
oospore—thick-walled, sexual spore of fungi in the class Oomycetes
ostiole (adj. ostiolate)—a pore through which spores are released from a fruiting body

paddy—a rice field
palea (pl. paleae)—the inner bract that, with the lemma, encloses the flower in grasses
panicle—an indeterminate inflorescence in which the main axis is branched, with pedicellate flowers borne on secondary branches
papillate—having a small, rounded protrusion
paraphysis (pl. paraphyses)—sterile hyphal elements that grow upward between asci in the ascocarp of fungi in the class Ascomycetes
parenchyma—plant tissue consisting of thin-walled cells functioning primarily in the storage and distribution of carbohydrates (the most abundant cell type in plants)
parthenogenesis—reproduction without sexual fertilization
pathovar—a subspecific group of individuals distinguished by pathogenicity to host plants of a certain genus or species only
perinuclear space—the space within the double-unit membrane of the nucleus of a cell
periplasmic space—the area between the cytoplasmic membrane and the cell wall in gram-negative bacteria; the area between the cytoplasmic membrane and the outer membrane in gram-positive bacteria
perithecium (pl. perithecia)—a subglobose or flask-shaped ascocarp
peritrichous—(of bacteria) having flagella distributed evenly over the surface of the cell
phenotype—the observable or manifest characters of an organism
phialide—a fungal conidiophore or cell of a conidiophore from which a basipetal succession of conidia develops through one or more openings
phloem—the principal food-conducting tissue of vascular plants
primary inoculum—inoculum causing the first or primary infection of a host, often being the survival form of the pathogen
promycelium—the germ tube of a teliospore, interpreted as a basidium
propagative virus—a virus that multiplies in its insect vector
pseudosclerotium (pl. pseudosclerotia)—a sclerotium that contains some host plant tissue
pycnidium (pl. pycnidia)—a more or less flask-shaped, ostiolate fungal structure lined with conidiophores
pyriform—pear-shaped

race—a subspecific group of individuals in a pathogen population distinguished by their ability to infect a certain set of cultivars of a given host
rachis—the main axis of an inflorescence
ratoon—a shoot growing from a plant that has been cut down
resistance—the ability of an organism to prevent damage from a pathogen or other factors
rhizosphere—the portion of the soil that encloses and is influenced by the roots of a plant
RNA—ribonucleic acid

saprophyte—an organism that lives on dead organic matter
sclerotium (pl. sclerotia)—a firm, often rounded, compact mass of dormant fungal mycelium that functions in survival
secondary inoculum—inoculum resulting from infection of a host earlier in the same growing season and functioning in the dispersal of the pathogen
seedborne—carried within or on seed
septate—divided into cells by cross walls (septa)
seta (pl. setae)—a stiff, usually pigmented, thick-walled, hairlike structure produced by some fungi
sheath—a more or less tubular structure surrounding a plant part or organ
sieve cell—a long, narrow sieve element typical of lower vascular plants that lack sieve plates
soilborne—carried in or present in soil
spikelet—a small or secondary spike; the basic unit of inflorescence in grasses, consisting of a short axis (the rachilla) and its two bracts (the glumes) and the florets and their bracts (the lemma and palea)
sporangiophore—a specialized fungal hypha that produce sporangia
sporangium (pl. sporangia)—a fungal organ that produces asexual spores internally or, in some cases, functions as a single spore
spore—the single- or multiple-celled reproductive unit in fungi, bacteria, and lower plants
sporidium (pl. sporidia)—a spore (basidiospore) produced from the promycelium (basidium) of smut or rust fungi
stamen—the pollen-producing structure of a flower
sterigma (pl. sterigmata)—a protuberance, usually conical, on which basidiospores are borne
stomate (pl. stomata)—a pore in the epidermis of a leaf or stem that allows for gaseous exchange between the plant and the environment
stroma (pl. stromata)—a mass of vegetative fungal hyphae, sometimes including host tissue or some other substrate, in or on which spores are produced
subglobose—not quite spherical
susceptibility—inability of an organism to prevent damage from a pathogen or other factors
symbiont—one of two dissimilar organisms living together in close

association, from which one, both, or neither of the organisms may derive benefit

sympodial—pertaining to the growth of a conidiophore from a succession of apices, each originating below and to the side of the previous apex

syncytium (pl. syncytia)—a multinucleate mass of protoplasm resulting from the fusion of protoplasts and surrounded by a common cell wall

systemic—distributed throughout the host rather than localized

teleomorph—the sexual state of a fungus

tiller—the shoot arising from a crown bud in grasses

tolerant—sustaining disease without serious damage or reduction in yield

transovarial—pertaining to the transmission of a parasite from one generation of a vector or host to the next through the egg

unitunicate—(of fungal asci) having only a single wall

vascular—pertaining to the conductive tissues (xylem and phloem) in plants

vector—an organism capable of transmitting a pathogen

vertical resistance—complete resistance that is effective against one or a few races of a pathogen

virion—an intact virus particle

viroplasm—a type of inclusion body formed at the site of virus replication and assembly in infected plant cells, where virions usually accumulate

virulence—degree of pathogenicity

viruliferous—pertaining to a vector that is carrying a virus

volunteer—a cultivated plant that has not been planted by human action

xylem—the water- and mineral-conducting tissue of vascular plants

zoospore—a motile, single-celled spore produced in a sporangium

Selected References

Little, J. R., and Jones, E. C. 1980. A Dictionary of Botany. Van Nostrand Reinhold, New York.

Singleton, P., and Sainsbury, D. 1987. Dictionary of Microbiology and Molecular Biology. 2nd ed. John Wiley & Sons, Chichester, England.

Swartz, D. 1971. Collegiate Dictionary of Botany. Ronald Press, New York.

Toothill, E., and Blackmore, S., eds. 1984. The Penguin Dictionary of Botany. Market House Books (Allen Lane, Penguin Books), Harmondsworth, England.

Index